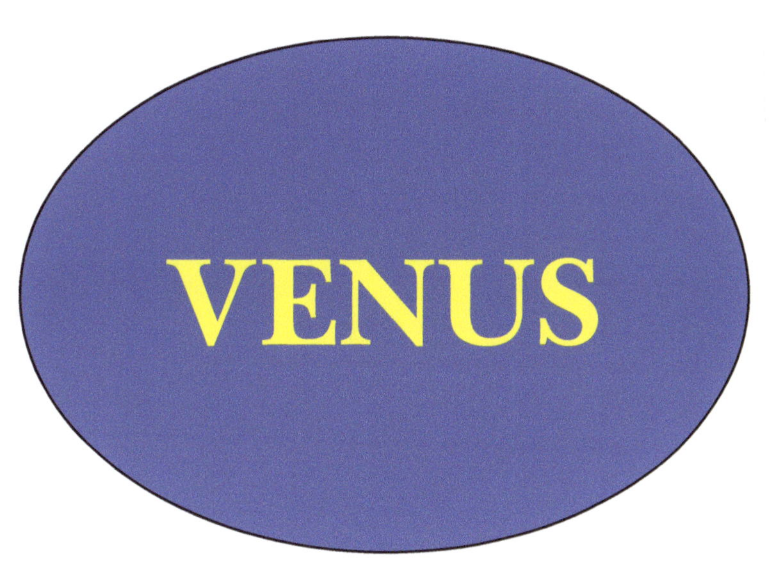

VENUS

A longer view
of history and of movements in the years ahead

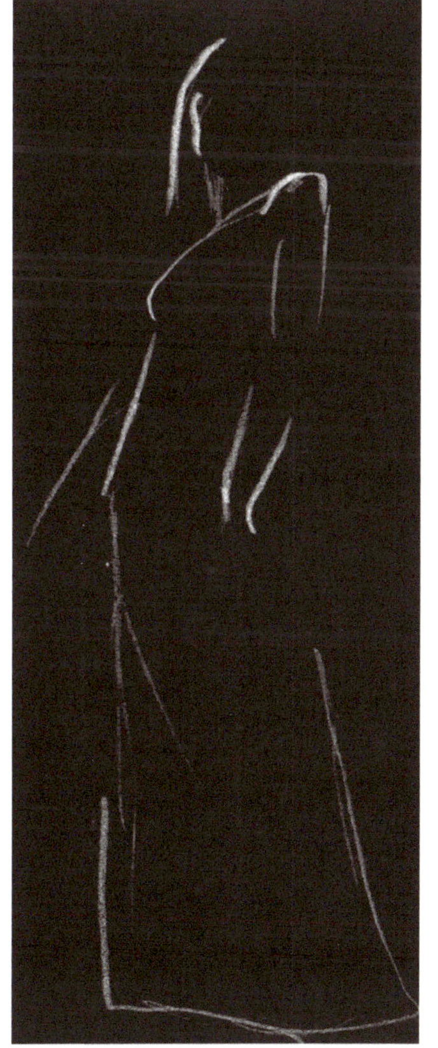

Goddess of love,
brightest planet,
sister of Earth

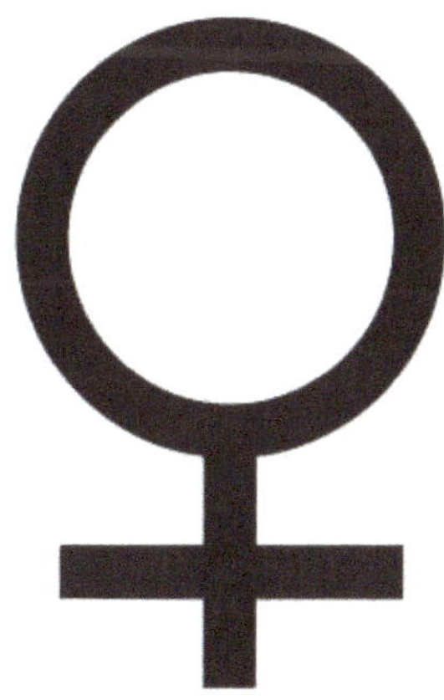

Guy Ottewell

F.R.A.S

ISBN 978-0-934546-81-2 revised 2020 August

To Barbara and Bonnie in the then and Tilly in the now

Front cover picture: Aphrodite, whom the Romans called Venus, was born from the foam, *aphros*, of the sea, and came surfing ashore on a scallop shell. My depiction of her owes something, probably apologies, to Botticelli and Velázquez (see the book's "Goddess" section). In the sky, you see her planet, vastly magnified, advancing into the constellation Taurus, between Orion and Perseus, and about to cross the star cluster of the Pleiades, as it did (see the "Pleiades Passages" section) in early April of 2020 and will do in each eighth April in the years ahead.

What looks like a broad crepuscular ray coming up from the horizon is actually a province of the sky: the region between the celestial equator (which cuts through Orion's belt) and the ecliptic.

Back cover picture: Venus showing its phases as it travels around its orbit. This is what we see from Earth, or rather from a viewpoint a few million miles above Earth so as to get a better view of Venus's orbit. Venus is shown at 10-day intervals, exaggerated 500 times in size (the Sun is exaggerated 5 times). The nearest Venus is at the moment called inferior conjunction, passing between Earth and Sun (as on 2020 June 3). The orbit appears tilted by 3.4°; the half to the left is slightly north of the ecliptic plane. The background stars are those of Orion, Taurus, and their neighboring constellations. Just above the Sun you can see red Aldebaran and the group called the Hyades.

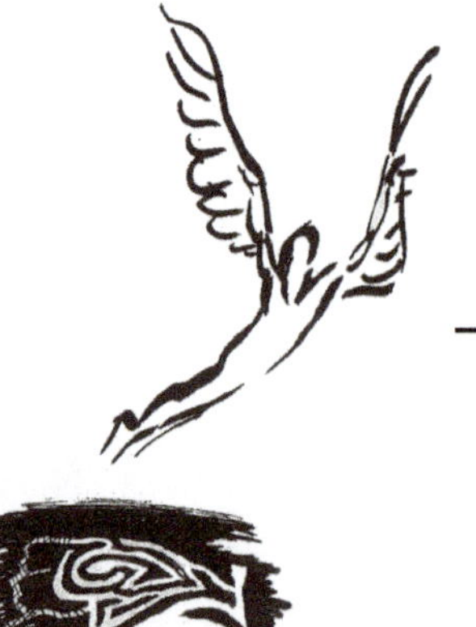

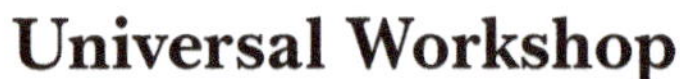

www.universalworkshop.com
Durham, N.C., U.S.A., and Isleworth, Middlesex, England

Preface

Each time I spent months preparing my annual atlas-sized *Astronomical Calendar* (of which the last and most elaborate issue was for 2016), I wished I could instead be creating charts for, say, Jupiter over a span of years instead of just one. So that is what I have begun to do. *Uranus, Neptune, Pluto: a longer view of their history and their movements in the years ahead* was published in 2018; and *Venus* is the second in the "Longer View" series.

My initial idea was just to provide illustrations and tables for spans of years—maybe the eight-year cycle of Venus, or the twelve or thirty years in which Jupiter and Saturn circle the sky. But as soon as you start saying something about a planet, you find that there are more and more things worth saying. Especially so for Venus.

I looked forward to getting fully into that eight-year cycle, and Venus's transits of the Sun and passages of the Pleiades. But I had not expected to find myself so interested by the varied stories of the space missions sent to or past Venus. And look for the section on "Physical Venus": Earth's "sister" planet is fundamentally different in startling ways.

And Venus is the planet of Love. You'll have to expect some enrichment of erotic color.

Don't forget that in this online age it's easy to find more information. For instance, one of my "fillers" quotes the last six lines of Andrew Marvell's "To His Coy Mistress." Using the title as search term, you can relish the cantering pace and brilliant choice of words in the whole of that high-pressure speech of seduction.

As a guide to observing, however, the core of the book is the section called "Apparition Pairs." It is tidier to describe the Venus cycle as five excursions into the evening sky and five excursions into the morning sky than as eight Earth-calendar years.

I intended to have *Venus* ready in good time for 2020, a fine year in the Venus cycle. And then at least for Valentine's Day. But the project burgeoned, as projects always do, also circumstances intervened. You now have before you a ladder of Venus years climaxing with 2028.

As with several previous books, I am heartily grateful to John Goss, Past President of the Astronomical League (www.astroleague.org), for proofreading and saving me from multiple typos and other flaws. And Alastair McBeath gave advice on the pages concerning ancient Mesopotamia. Any remaining errors will have been caused subsequently by me.

\acom\gods.eps

Simplified family tree of the Greek gods (Roman names in parentheses, major planet names bold)

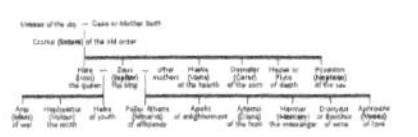

Contents

10

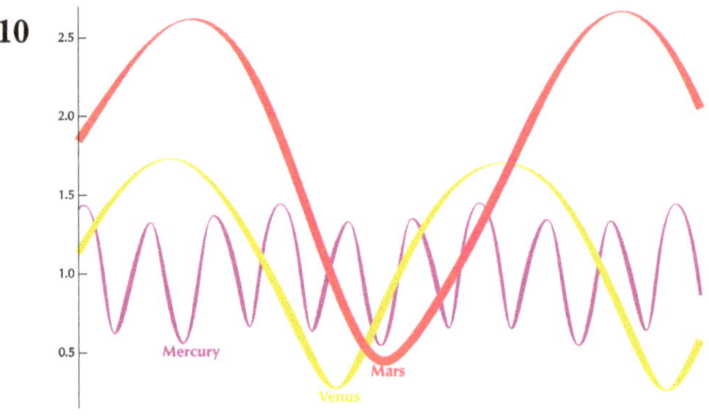

15

east (evening)
Sun
west (morning)

26

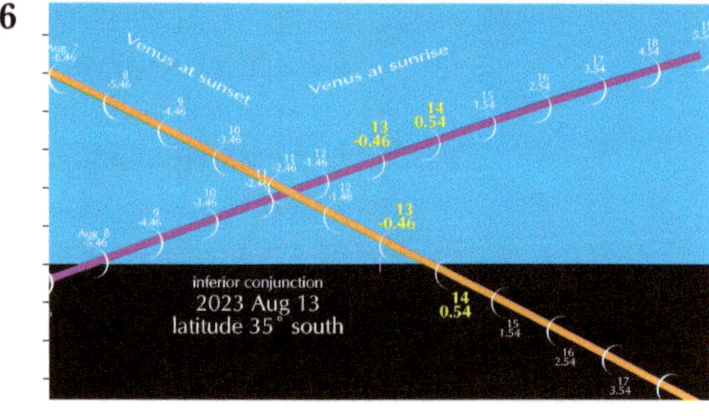

31

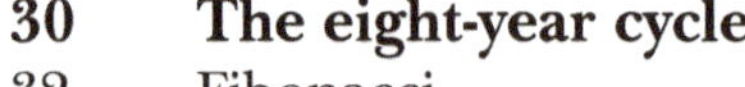

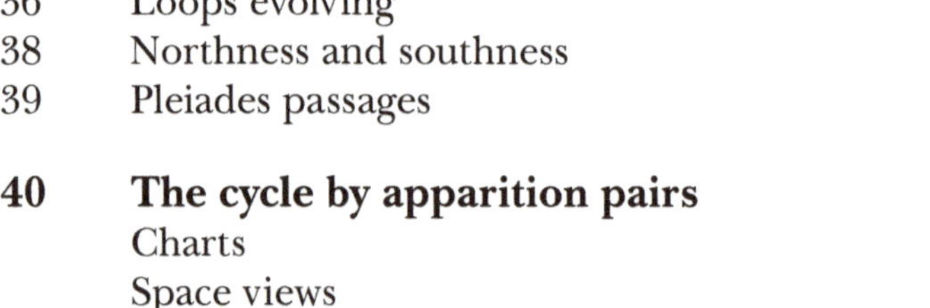

39

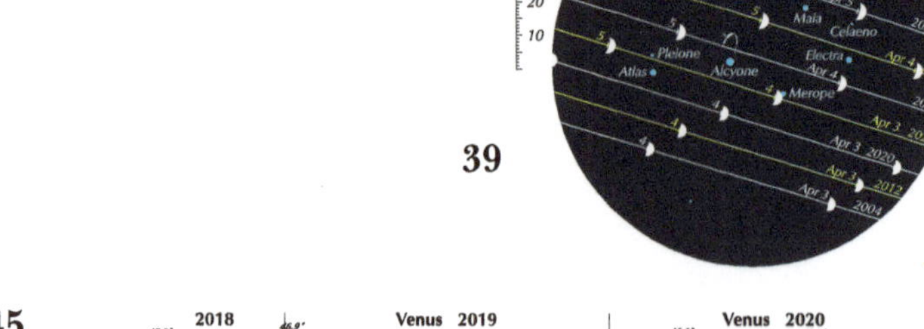

45

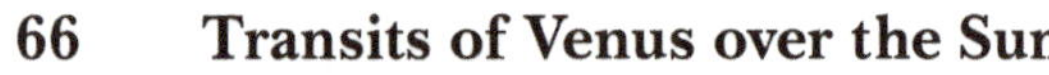

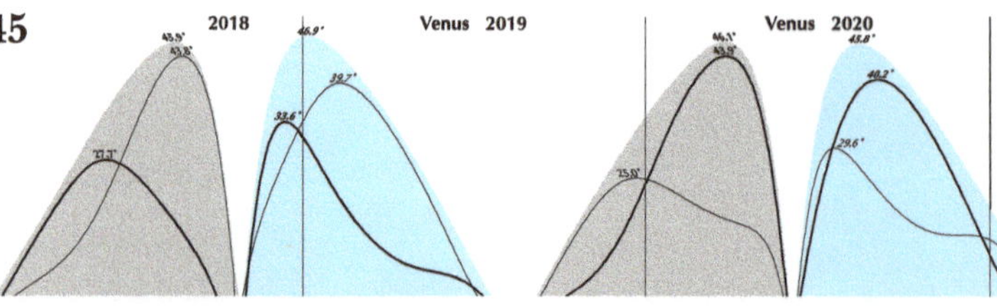

69

75

77

88

89

102

114

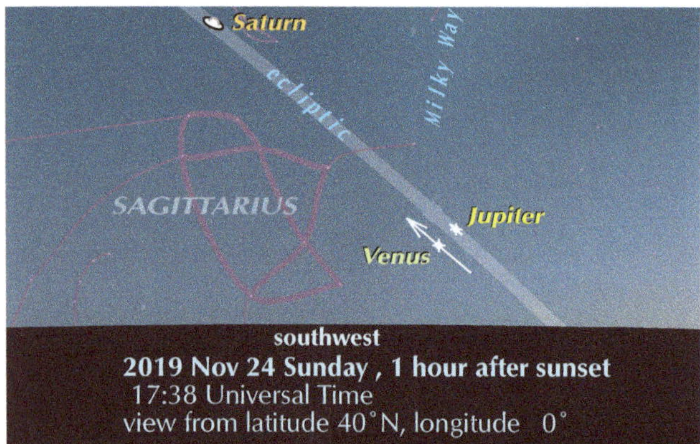

Venus passing Jupiter on 2019 Nov. 24, as seen after sunset from a location in Earth's northern hemisphere. At the time, the brightnesses of Venus, Jupiter, and Saturn, expressed in the astronomical system of "magnitude," were -3.9, -1.8, and 0.6, which meant that Venus was about 7 times brighter than Jupiter. Venus was to pass Saturn on Dec. 11. The arrow through the star symbol for Venus shows its motion over 5 days.

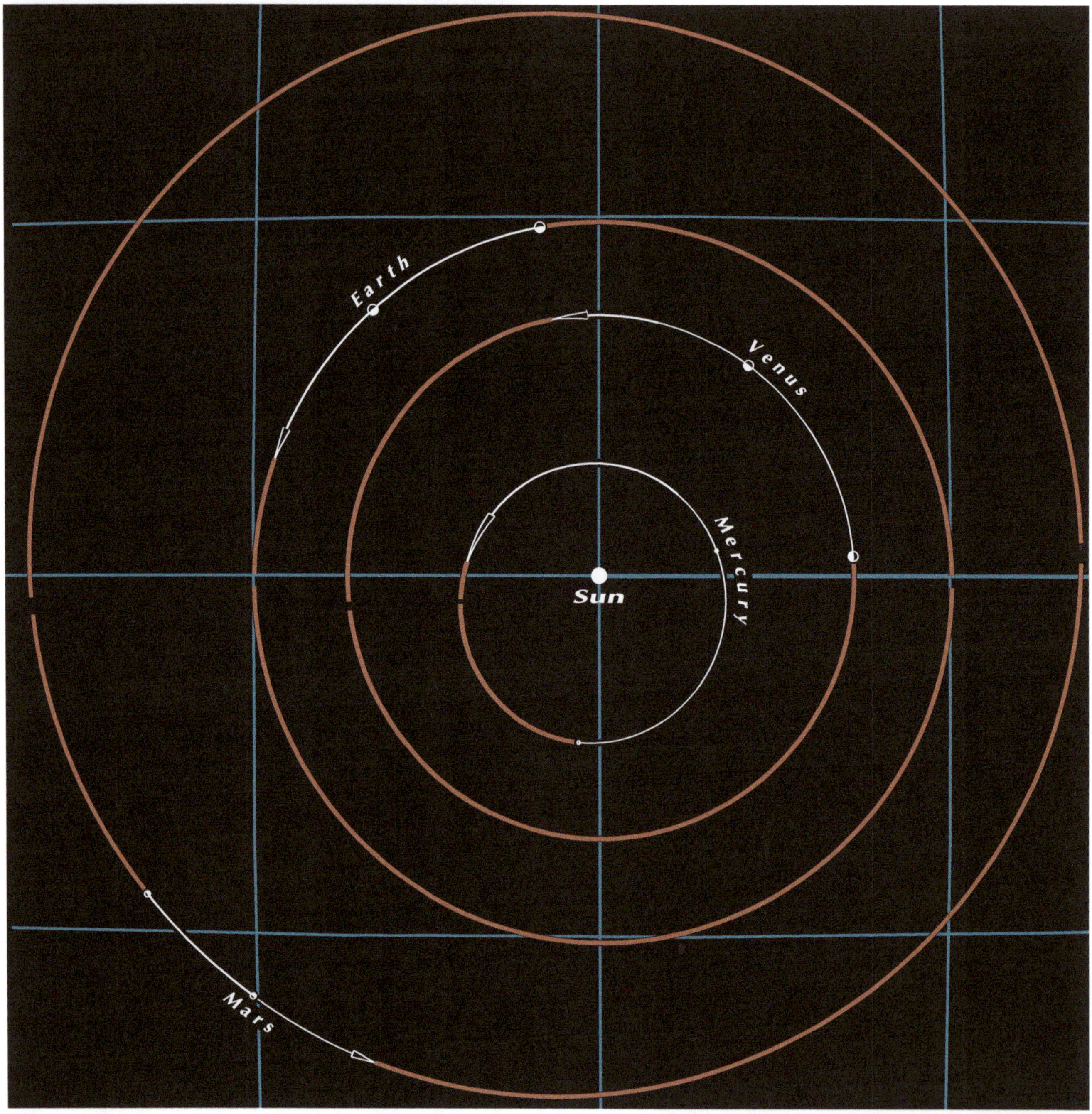

Orbits of the four inner or terrestrial ("Earth-like") planets, seen from ecliptic north. The planets' courses are shown over the first two months of 2020. Their globes, at the beginning of each month, are exaggerated 300 times in size, the Sun's 5 times. Lines on the ecliptic plane are drawn at intervals of 1 astronomical unit, the average distance from Sun to Earth. The orbits are not circular but elliptical, especially those of Mercury and Mars. Venus is about half as far out as Mars. The vernal equinox direction is to the right.

Star, goddess, planet

Venus is the brightest "star." It shines more brightly for us than any actual star except the Sun, and than any other celestial body except the Moon and a very few brilliant comets.

It was, surely, the first "star," or heavenly point of light, that was perceived to change position in relation to the other stars, and so the first to be distinguished as a "planet"–a wanderer. We can guess, though we cannot be certain, that ancient people noticed its movements before they noticed those of Jupiter or Mars. I would guess that the awareness goes back about as far as the beginnings of language.

Venus is the planet that comes nearest to Earth, and is Earth's almost-twin sister in size, though very different in constitution.

She or It?

Planet Venus has long been associated with the goddess of love and maternity, and so is often referred to as "she."

The ancients conceived of all planets as personalities, therefore with gender. Upholding the anthropomorphizing tradition is pleasant–up to a point. It makes the celestial bodies feel like friends, and reminds us of the origins of their names and the history of ideas associated with them. With apology to those who prefer this habit, I shall not maintain it, because it comes to seem a tiresome affectation if repeated, as would be necessary here.

We can allude to her dance around the Sun, and to her associated goddess, but not all the time.

Albedo, brightness, magnitude

Planets do not emit light as stars do, but reflect the Sun's light. Venus is light in color. Its nearness to us, and size, make it appear brighter than any star, but it is rendered brighter still by its reflective surface.

The measure of this is *albedo,* a word that in Latin means "whiteness" but in science is defined as the fraction of incident light that a body reflects. Albedo 1 would mean that all light is reflected. Venus's albedo is 0.72. By comparison, Earth is little more than half as relatively bright, with albedo 0.39. Venus's great "whiteness" is due to clouds, with which it is entirely wrapped, whereas on Earth they leave much of the dark ocean showing.

The astronomical way of expressing brightness is *magnitude,* originally a number representing order, rather than a quantity. "First-magnitude" stars have magnitudes around 1, the dimmest visible to the naked eye are at about 5 or 6. The very brightest objects have zero or even negative magnitude: Sirius, the brightest star, -1.4, the Full Moon about -13, the Sun -27. The scale is logarithmic: one magnitude step means a factor of about 2.5 in quantity of light.

Venus at its peaks of brightness reaches a magnitude just under -5. This means that Venus then sends us about 25 times as much light as Sirius.

Daylight visibility

Venus is bright enough to be seen in clear sky when the Sun is above the horizon. It is often noticed not long before sunset, or after sunrise, if fairly high in the sky. People have been known to exclaim "What's that?" and take it for an Unidentified Flying Object.

At total eclipses of the Sun, as the sky darkens, Venus is the first "star" to be seen (look with caution!), as much as thirty minutes before totality begins. Only after that is the sky dark enough to reveal a few of the next brightest planets and stars.

With optical aid, Venus can be found when the Sun is high, if you know where to look. If it is not far from the Moon, that can be used as a guide to it.

Indeed telescope-users prefer to observe Venus by day or in bright twilight, because its contrast with a dark background can make it too dazzling.

Venus at midnight

Venus, then, may be seen not only throughout daylight but quite far into the night. But not, of course, in the middle of it. Since the orbit lies between ours and the Sun, Venus must always be in the sunward half of the sky, if sometimes near to the limit of it.

But there is almost an exception to this. At some times and places, such as the northern hemisphere in April of 2012, or May and June of 2015, Venus sets little more than an hour before midnight. By the artificial "Daylight-Saving Time," better called Daylight-Shifting Time, 11 o'clock is called "12," so Venus is setting essentially at midnight, an obvious physical impossibility.

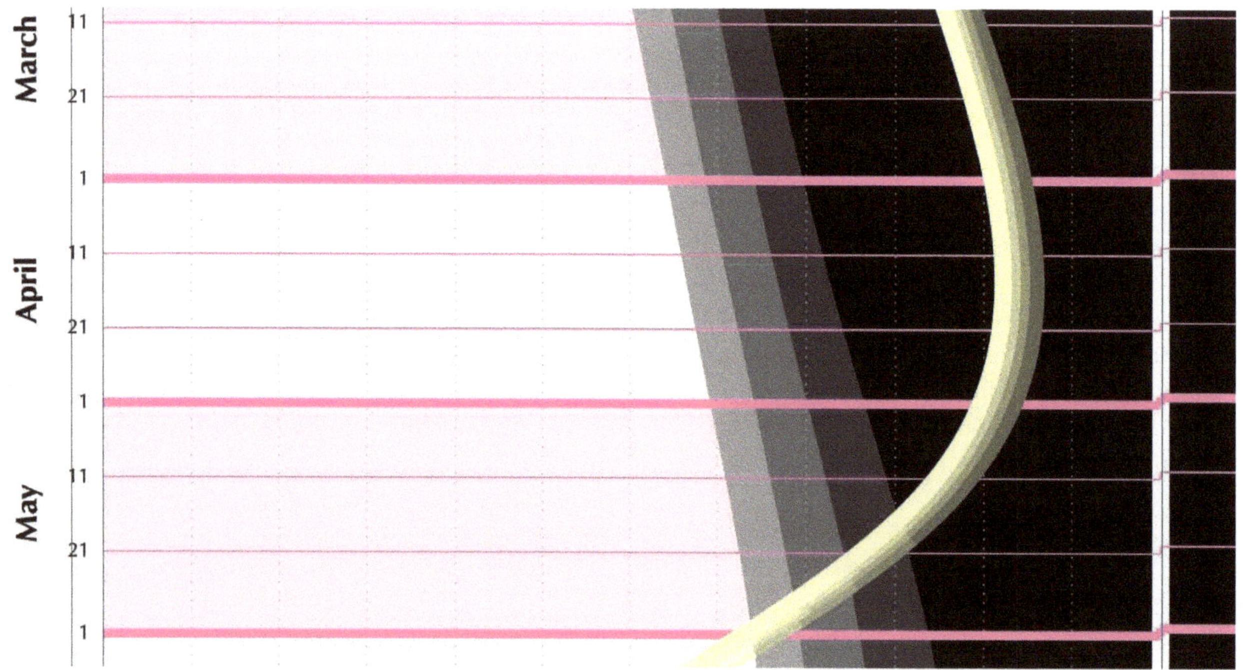

In this graph, adapted from part of the "Rising and Setting" graph in my Astronomical Calendar for 2012, the vertical white line represents midnight, and the dotted lines are at the other hours. The outer, darker side of the curve represents the time when Venus set, for latitude 40° north. By summer Daylight Saving Time, clocks were an hour ahead of natural time. So the 11 p.m. line was "midnight." Venus came close to touching that time. For latitudes farther north, it would have come even closer to touching.

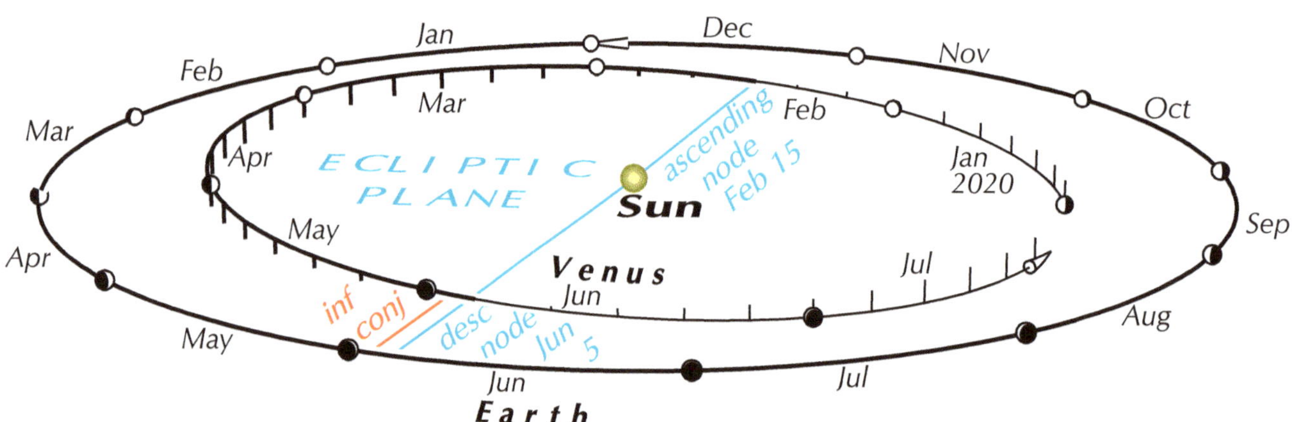

Spatial view showing the orbits of Earth in a year and speedier Venus over the first 7 months of 2020. Earth moves in the ecliptic plane; Venus in a plane inclined 3.4° to the ecliptic. The planets, at the beginnings of the months, are exaggerated 300 times in size and the Sun 5 times. Stalks connect Venus to the ecliptic plane at intervals of 5 days. Also, Venus's path is drawn with a thinner line when it is south of the ecliptic. In 2020 it happens that descending node almost coincides with inferior conjunction.

Heliocentric motion

That is, movement relative to the Sun.

Of the eight major planets, the first two in order outward, Mercury and Venus, are classified as the "inferior" planets. That doesn't mean they are of poor quality. The literal meaning of the Latin word *inferior* is "lower," and Mercury and Venus are indeed deeper down than us in the gravitational field of the Sun. Mars, Jupiter, and the rest are the "superior" planets, higher out from the Sun.

Distances in the solar system are reckoned in terms of the astronomical unit (a.u.), the Sun-Earth distance of about 93 million miles or 150 million kilometers. Venus's average distance from the Sun is 0.72 a.u.–somewhat less than three-quarters of Earth's.

So the distance from our orbit inward to Venus's is about 0.28 a.u. And the greatest distance from us to Venus, when it is on the far side of its orbit, is about 1.72 a.u.

Nearly circular orbit

The orbits of the planets are not perfect circles, but ellipses. Venus's is the nearest to being circular. Its eccentricity is 0.007, which means, roughly, that its greatest and smallest distances from the Sun (aphelion and perihelion) differ by that amount from the average distance. They are about 0.728 and 0.718 a.u.

The orbit next closest to circular is Neptune's, with eccentricity 0.009. Earth's is 0.017. The most strongly elliptical planetary orbits are those of Mercury, with eccentricity 0.206; Mars, 0.093; and Pluto–no longer considered a major planet–0.25.

Love makes the world go round

–1935, words by Edward Heyman, music by Dana Suesse. There have been at least 10 versions since, including in the musical Carnival!, 1961.

An English boy on a school trip to France asked the chambermaid in the hotel at Menton not to clean his room yet but to come back "in a minute." Or intended to, but she came back at midnight. *Minute, minuit.* He was too young to take advantage.

Plane, inclination, nodes

Earth travels around the Sun in a plane, called the ecliptic plane. This is mapped as an imaginary line, the ecliptic, around the sky. The ecliptic is like a path, along which the solar-system bodies appear to travel, because they orbit in planes only a little different. At least, the Sun, Moon, and major planets stay close to the path; many minor bodies, especially comets, stray farther from it.

Venus orbits in a plane that slopes across the ecliptic at an angle of about 3.4°. This is Venus's orbital inclination.

The angle may seem quite small, but is actually the second largest for the major planets: Mercury 7°, Saturn 2.5°, Mars 1.9°, Neptune 1.8°, Jupiter 1.3°, Uranus 0.77°. (Pluto, not a major planet, has a large orbital inclination, 17°.)

Moreover, Venus, because of the nearness of its orbit to ours, can appear considerably more than 3.4° north or south of the ecliptic. This happens around the time when it comes nearest to us, that is, inferior conjunction. It reaches as much as 8.33° north of the ecliptic, and 8.26° south of it.

Venus's orbital plane cuts northward through the ecliptic at about longitude 257°; which means that, seen from the Sun, it is in the direction of the constellation Ophiuchus–just north of the great red star Antares in Scorpius.

The opposite point, the descending node, where Venus during every orbit slides southward through the ecliptic, is at longitude 77°. This is in Taurus, not far northeast of the other reddish first-magnitude star, Aldebaran.

By "node" we can mean this point in the map of the sky; or the geometrical point on Venus's orbit; or a time, the instant at which Venus reaches this point.

The planet that comes nearest

Venus can come nearer to us than any other planet can, because its orbit is nearest to ours. It is nearer than that of Mars, the next planet outward.

That doesn't mean that Venus is at all times our nearest neighbor. Two planets, to be at their nearest to each other, have to be in the side-by-side parts of their orbits, so Venus is near when it is approximately between us and the Sun (inferior conjunction) and Mars is near when it is approximately straight outward from the Sun (opposition).

There are stretches of time when Mars is nearer to us than Venus is, but they are relatively short. And Mercury, surprisingly, though it is two orbits away from us, and cannot come as near as either Venus or Mars, is our nearest neighbor for the longest stretches of time.

I found that, over a sample period of 1,000 days, Mercury was the nearest to us for 529 days, Venus for 295, and Mars for 176.

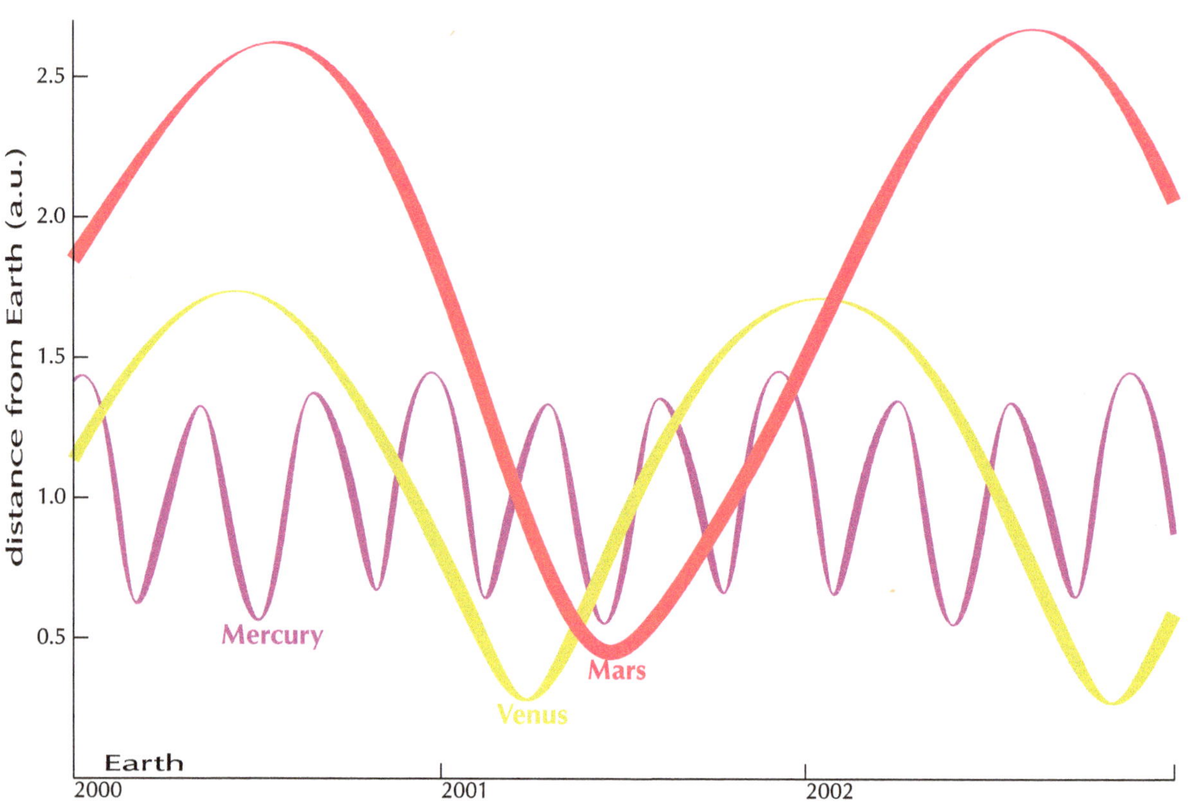

The distances of Mercury, Venus, and Mars from Earth, in astronomical units (Sun-Earth distances), over a span of three years. The curves are thicker when the planets more observable because at greater elongation from the Sun.

Seven and a half months

Venus goes around the Sun in 225 days: about 7½ months), or 0.62 of an Earth-year. This is its *sidereal period*: the time it takes to revolve back to the same direction in space–the same position against the starry background, as seen from the Sun.

Distance and period are tied to each other by Kepler's third law of planetary motion: period squared equals mean distance cubed.

The period was known long before the distance, because position in the sky can be directly observed. Distance cannot. In fact all distances, not just Venus's, were unknown until Venus itself enabled us to find them–a whole other subject, that of transits, to which we will come.

Venus's period, more accurately 0.61519 year, squared is 0.37845; the cube root of that is 0.72333 of an astronomical unit. The problem that remained was to find the astronomical unit.

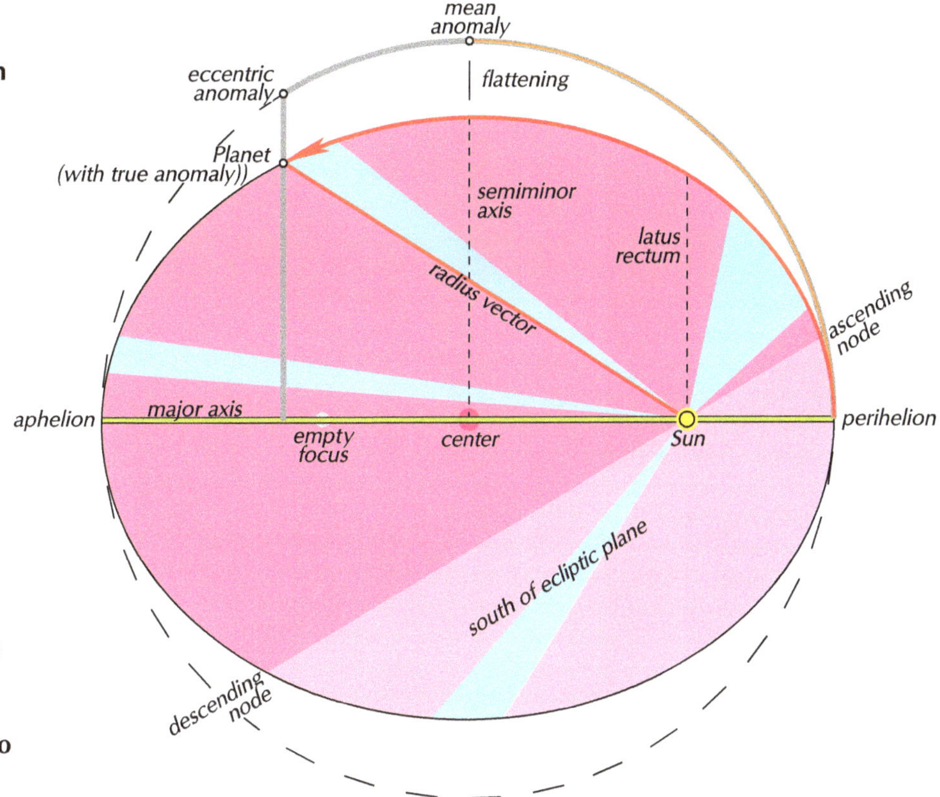

An orbit, illustrating Kepler's laws of planetary motion. Johannes Kepler published them in 1609 (the first two) and 1619.

(1) The planet moves in an ellipse, with the Sun at one of the two foci of the ellipse.

(2) The radius vector (line from Sun to planet) sweeps out equal areas in equal times. (Each sample blue area is covered in 1/32 of the period.)

(3) The semimajor axis or mean distance (in astronomical units) is equal to the cube root of the period (in years) squared.

"Anomaly" here means angular distance traveled from the perihelion. Mean anomaly and eccentric anomaly are steps used toward calculating the true anomaly, which is found by Kepler's transcendental equation.

The semimajor axis is equal to the mean distance from the Sun (the mean radius vector). The flattening is actually the proportion between the length shown and the semimajor axis. The eccentricity has a more difficult formula: 1 minus (perihelion distance divided by semimajor axis).

Galileo and the Mother of Loves

The telescope was invented by Dutch spectacle-makers in 1608. Galileo Galilei, a professor of mathematics at Padua, learned of it in 1609, made his own, and applied it to astronomy. His discoveries confirmed the heliocentric thinking of Copernicus and Kepler, which upset the dogma of a cosmos circling a stationary Earth.

The clincher was Galileo's noticing, in December 1610, the changing phases of Venus. They proved—almost shouted—that Venus orbits the Sun. To Kepler in Prague, and other scholars, he sent by stagecoach mail a Latin riddle:

HAEC IMMATURA A ME IAM FRUSTRA LEGUNTUR O Y

– "These unready things are as yet read by me in vain O Y." It was an anagram, encoding the discovery so that he could claim priority if it proved true. Kepler, solver of deep problems, had ingeniously mis-solved a previous anagram from Galileo (discovery of Saturn's two bulges) to mean that Mars had two moons—which it does, but they were not discovered till 1877! This time, flummoxed, he wrote back: "I beg you not to leave us long in doubt." Galileo watched Venus till he was sure about those phases. On the first day of 1611 he sent Kepler the solution:

CYHNTHIAE FIGURAS AEMULATUR MATER AMORUM

– "The Mother of Loves imitates the shapes of Cynthia." Not so mysterious, if like all European gentry you had a classical education: "Venus shows phases like the Moon." "Cynthia" was an epithet of Moon goddess Diana, born on Mount Cynthus in the island of Delos.

The line is a dactylic hexameter, slightly disobeying strict scansion: the *i* of *Cynthiae* is short instead of long, the *ae* of *aemulatur* is long instead of short. Those who wrote in "the ancient and marmoreal tongue" (Helen Waddell) had become vague about its phonetics.

As for "O Y," perhaps they were a hint that it was anagram, into which they could not fit. But we picture Galileo exclaiming like a town crier or a jazzman: "HAEC IMMATURA A ME IAM FRUSTRA LEGUNTU OH YEA!"

He had reason to use cautious code. He was not burned at the stake like Giordano Bruno, but had to spend his remaining years in house arrest. He was cleared of error by the pope in 1992.

Geocentric motion

That is, behavior as seen from Earth.

Venus goes around the Sun in 225 days (its sidereal period), while slower Earth goes around its more outward track in 365 days, so Venus keeps overtaking us on the inside. But when it comes around to the point where it last overtook us, we have moved on, by about another 200° of the circuit. So the time Venus takes to overtake us is longer: 584 days.

This is its *synodic period*. It is the average period between Venus's Earth-related events, such as from one passage between Earth and Sun and the next. In that time it travels around the circle of the sky about 584/360 or 1.6 times.

"Synodic" is an abstruse term, but it's hard to think of a better; it means the repeating sequence of Venus performances. It is a product of the movements of Venus and Earth. It happens to make a rhythm: you will notice that events in it recur after eight years. We'll dwell later on this eight-year cycle.

Apparitions: the Evening Star and the Morning Star

We see Venus some of the time on one side of the Sun, then on the other. These are its alternate "apparitions," or appearances.

From superior conjunction–the position behind the Sun–Venus comes out on the Sun's east side. This is the beginning of an evening or eastern apparition. Moving faster than Earth, it circles around and overtakes us on the inside, passing between us and the Sun at inferior conjunction. From there, it moves out on the western or morning side. And it swings on around to the next superior conjunction beyond the Sun.

The way this appears to us is that at superior conjunction an evening apparition begins. Venus appears higher in the sunset sky each evening, setting longer after the Sun. It reaches a greatest eastern elongation, or angular distance from the Sun. Then it appears to drop rapidly lower in the sunset sky, disappearing around the time of inferior conjunction. Then it reappears, rising before the Sun and beginning its morning apparition, and moves out toward a greatest western elongation.

The lengths of these apparitions have to average 292 days (half of Venus's synodic period of 584), but they vary between under 287 and over 296, because each contains more or less of the time when Venus is in the slightly swifter half of its orbit.

It is far from obvious that the supreme star that hovers for many evenings over the western horizon is the other that is seen in mornings, in the opposite direction. When ancient people realized that the two are one, it could well have seemed to travel from west to east through an abyss under the flat Earth.

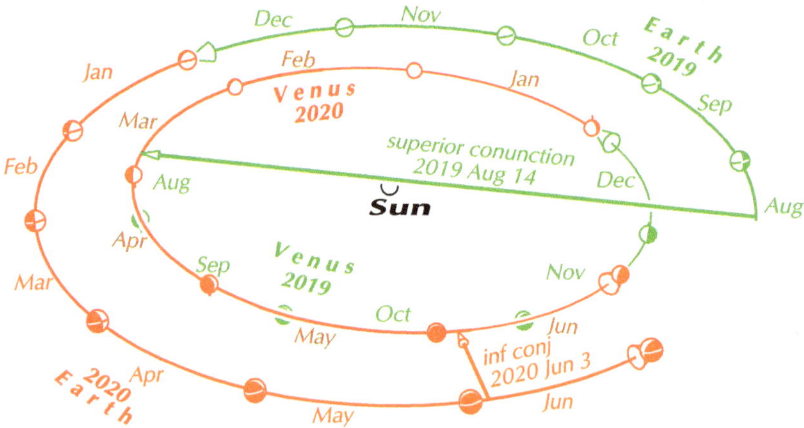

The path of Venus during one apparition, and of Earth during the same time, with sightlines from Earth to Venus at the beginning and end dates of the apparition. This is the evening apparition from the superior conjunction of 2019 Aug. 14 to the inferior conjunction of 2020 June 3. Over about the last 3 months of the nearly 10-month apparition, Venus overlaps its earlier course. To try to make this clear, the paths are drawn with dull green in 2019, when Venus is relatively distant and dim, and red as it becomes nearer and brighter in 2020. This spatial view is from 40° north of the ecliptic plane, at ecliptic longitude 240°.

Hei quam felix transitus
amoris in soporem,
sed suavior regressus
ad amorem!

Hei! how good the slide
From love to sleep,
But sweeter far the glide
From sleep to love!

—Part of the long poem that begins "Dum Dianae vitrea sero lampas oritur" ("When Diana's crystal lantern rises late") in the Carmina Burana, "songs of Buria." This 13th-century collection of 254 Latin, German, and Provençal poems and other fragments, often satirical or bawdy, was discovered in 1803 in the monastery at Benediktbeuern in Bavaria. They were mostly the work of the drifters called Goliards: students and junior clergy who found lodging in monasteries or wherever they could. The "Dum Dianae" was called by Helen Waddell "the summit of mediaeval secular poetry as the 'Dies Irae' is of sacred." It ends with "Veneris milicia"—the "troops of Venus," who fluctuate between hope and despair.

Dulcissime!
Totam tibi subdo me!

—Another fragment from the Carmina Burana, set in Carl Orff's music as a high clear soprano scream. There is no way English can do what Latin does with the feminine inflexion -am and the (perhaps invented) verb subdo.

Longitudes

The great differences between the apparitions are caused by the relative positions of Venus and Earth—that is both obvious and unclear. We can pin these differences down by considering ecliptic longitudes, which are positions around the circle of the sky, measured from the vernal equinox point in Pisces. Each apparition begins at a certain longitude, so that fixes the longitudes at which its other events appear.

At the superior conjunction (2019 Aug. 14) that begins the 2019-2020 evening apparition, Venus is at longitude 141°, and, from our own position at that time, Venus is entering Leo. Over the next seven and a half months, it pulls out to easternmost elongation (2020 Mar. 24) almost back to the same longitude, at 140°, which then, because we have traveled around too, results in our seeing it in Aries. And it comes to inferior conjunction, the end of that apparition, on 2020 June 3 at longitude 254°, appearing to us in Taurus. All the other effects at the various times, such as its height above or below the ecliptic, result from where it is in its orbit, that is, its longitude.

But at the beginning of the next evening apparition (2021 Mar. 26), it is at longitude 6°, in Pisces—224° farther on. So the events of that apparition happen at longitudes farther on.

In general, each apparition has slewed around forward from the previous one by an average of 3/5 of the circle, or 216°. (It's an average, varying from about 203° to 225°.) That, after the cycle of five, comes back to the same longitude for the start of the next cycle.

You can clearly see this slewing around of the apparitions in our spatial diagrams of them.

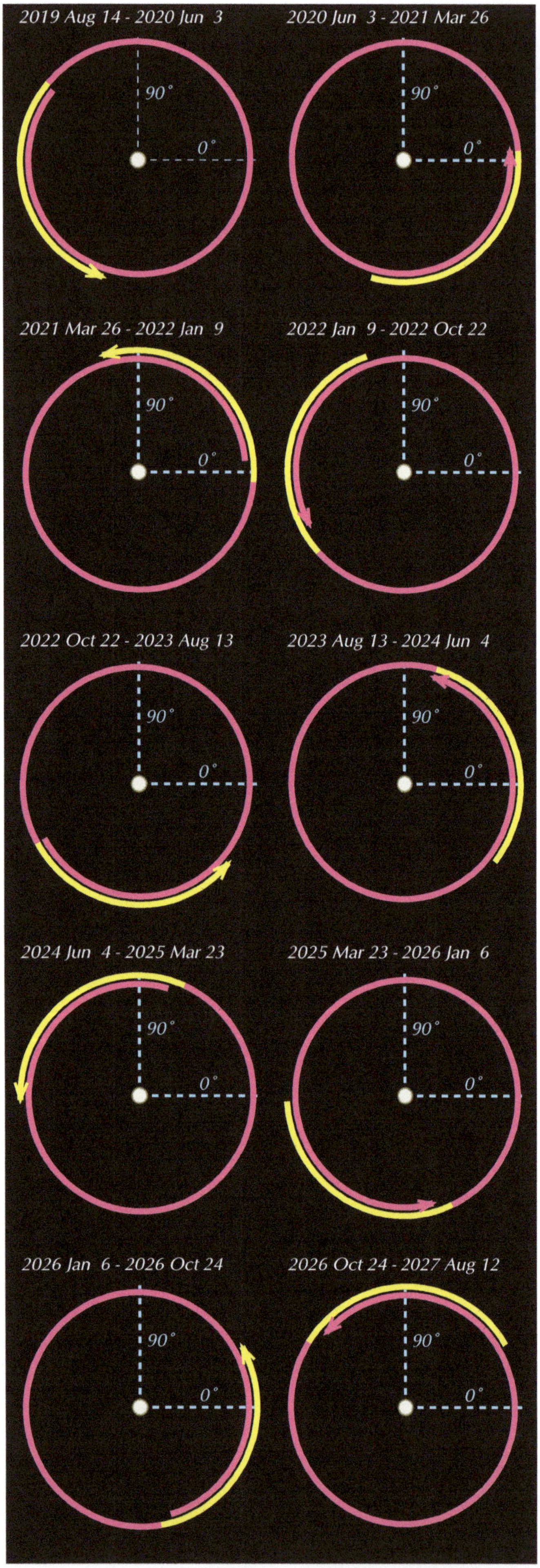

2019 Aug 14 - 2020 Jun 3

2020 Jun 3 - 2021 Mar 26

2021 Mar 26 - 2022 Jan 9

2022 Jan 9 - 2022 Oct 22

2022 Oct 22 - 2023 Aug 13

2023 Aug 13 - 2024 Jun 4

2024 Jun 4 - 2025 Mar 23

2025 Mar 23 - 2026 Jan 6

2026 Jan 6 - 2026 Oct 24

2026 Oct 24 - 2027 Aug 12

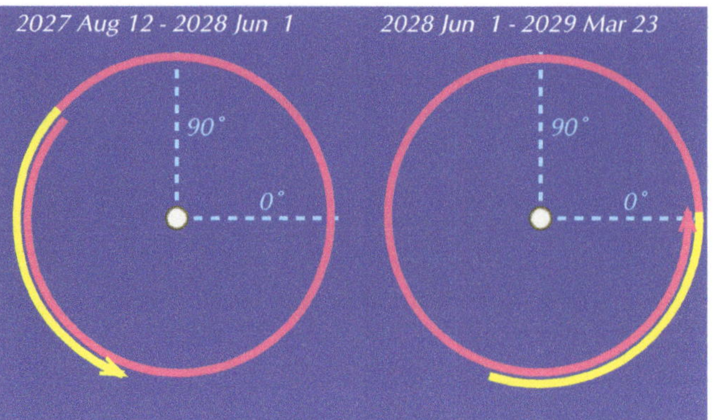

Schematic diagrams of Venus going around the Sun. Each diagram represents Venus's heliocentric track during one apparition. In the left column are evening apparitions, starting at superior conjunction and ending at inferior conjunction; in the right column, morning apparitions, from inferior back to superior conjunction. Distance from the Sun is manipulated so as to show how the track overlaps itself.

In yellow is the part of the track from easternmost elongation (in the evening sky) to westernmost elongation (in the morning sky). Let your eye go from the yellow arrow-tip (inferior conjunction) to the resumption of yellow in the diagram to the right. The yellow parts represent the rush between Earth and Sun.

Each apparition track begins where the previous one ended. After 5 pairs of apparitions, covering 8 years, superior conjunction is back almost exactly to where it was at the beginning; the cycle begins again with a near-repetition of the first pair.

The dashed lines from the Sun are at longitude 0° (in the Pisces direction) and 90° (Taurus-Gemini).

O waly, waly, but love be bonnie
A little time while it is new,
But when 'tis auld it waxeth cauld
And fades away like morning dew

– "The Water is Wide," Scottish folk song with a haunting tune

"I long woo'd your daughter, my suit you denied;
Love swells like the Solway, but ebbs like its tide.
And now am I come, with this lost love of mine,
To lead but one measure, drink one cup of wine."

No, he has come to snatch her away on horseback from her forced wedding to "a laggard in love and a dastard in war." The "Young Lochinvar" episode in Canto V of Sir Walter Scott's *Marmion*, 1808.

Events in the apparitions

The conjunctions with the Sun

In a way, the two most important of the geocentric or Earth-centered events are the two that essentially can't be seen from Earth: the superior conjunction, beyond the Sun, and the inferior conjunction, between us and the Sun. For these are the moments that begin and end apparitions, the moment when the planet transfers from one half of the sky to the other.

"Inferior" again makes sense, as it does for classifiying Venus as an "inferior" plant: it is at this moment down between us and the Sun. It is like a bird flying between an observer at a high window (us) and the ground (the heavy Sun).

"Superior" doesn't fit quite so well. It is apt only in being the opposite of "inferior"; but Venus, over there in the distance, is still lower toward the Sun than we are.

The superior planets–those farther out than us–cannot venture between us and the Sun and so have no inferior conjunctions; their Sun-conjunctions are all superior, and are just called conjunctions.

Waxing and waning

At superior conjunction, Venus, if we could see it, would be face on to us, a complete illuminated circle like the full Moon, but at its smallest because it is at its most distant.

As it comes around its circle toward overtaking us, it appears in the evening sky to the east of the Sun (left, as seen from our northern hemisphere), setting increasingly later after the Sun. It grows, in binoculars or telescope, larger, less full (more and more of its dark side turned toward us) but with greater total brightness. Coming nearest, it falls back again toward the Sun, becomes a crescent, ever slenderer but longer, until, in front or almost in front of the Sun, it offers us only a black circular face, at maximum diameter–still very small compared with the Sun.

As it reappears in the morning sky and rushes away ahead of us, it goes through these transformations in reverse order, appearing smaller, rounder, dimmer.

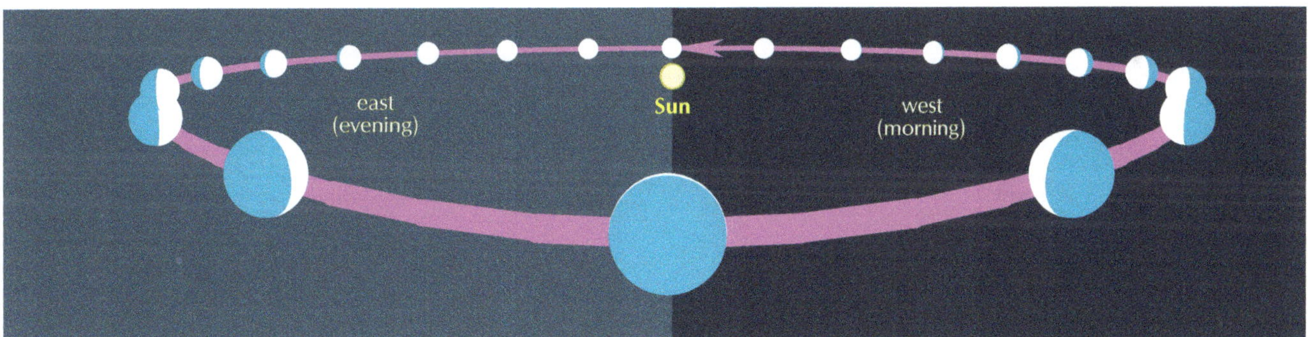

The changing phases of Venus. The planet is enormously exaggerated in size. The view is like that from the Earth, but from 0.1 astronomical unit of it. The lighter and darker colors of the background indicate the evening and morning halves of the sky.

But silken nets and traps of adamant will Oothoon spread,
And catch for thee girls of mild silver, or of furious gold;
I'll lie beside thee on a bank & view their wanton play
In lovely copulation bliss on bliss with Theotormon;
Red as the rosy morning, lustful as the first born beam,
Oothoon shall view his dear delight, nor e'er with jealous cloud
Come in the heavens of generous love; nor selfish blightings bring.

–A stanza near the end of "Visions of the Daughters of Albion," 1793, one of William Blake's privately printed poems embodying his personal mythology. In it, he condemned enforced chastity and marriage without love.

Elongation climaxes

After passing behind the Sun (superior conjunction), Venus swings out into our evening sky, until it reaches a greatest elongation–that is, apparent or angular distance from the Sun.

Then it comes around in front of us (inferior conjunction) and climbs out into our morning sky, until it reaches an elongation maximum in that westward direction.

At each of the moments of greatest elongation, a line from us to Venus is tangent to Venus's orbit.

The greatest elongations average about 46°. More precisely, because of the slight ellipticity of Venus's orbit, and our varying viewpoint in our own slightly elliptical orbit, the amount varies, from 45.38° to 47.32°.

This is in contrast with the other inner planet, Mercury, whose orbit is more strongly elliptical. Mercury comes out from the Sun only about half as far as Venus does, and its elongation peaks vary widely, from 17.87° to 27.83°.

Venus, then, can leap out from the Sun to a distance that is more than half of a right angle. This means that it can appear not only throughout the time of twilight but in full darkness, setting up to 3½ hours after the Sun, or rising as long before the Sun.

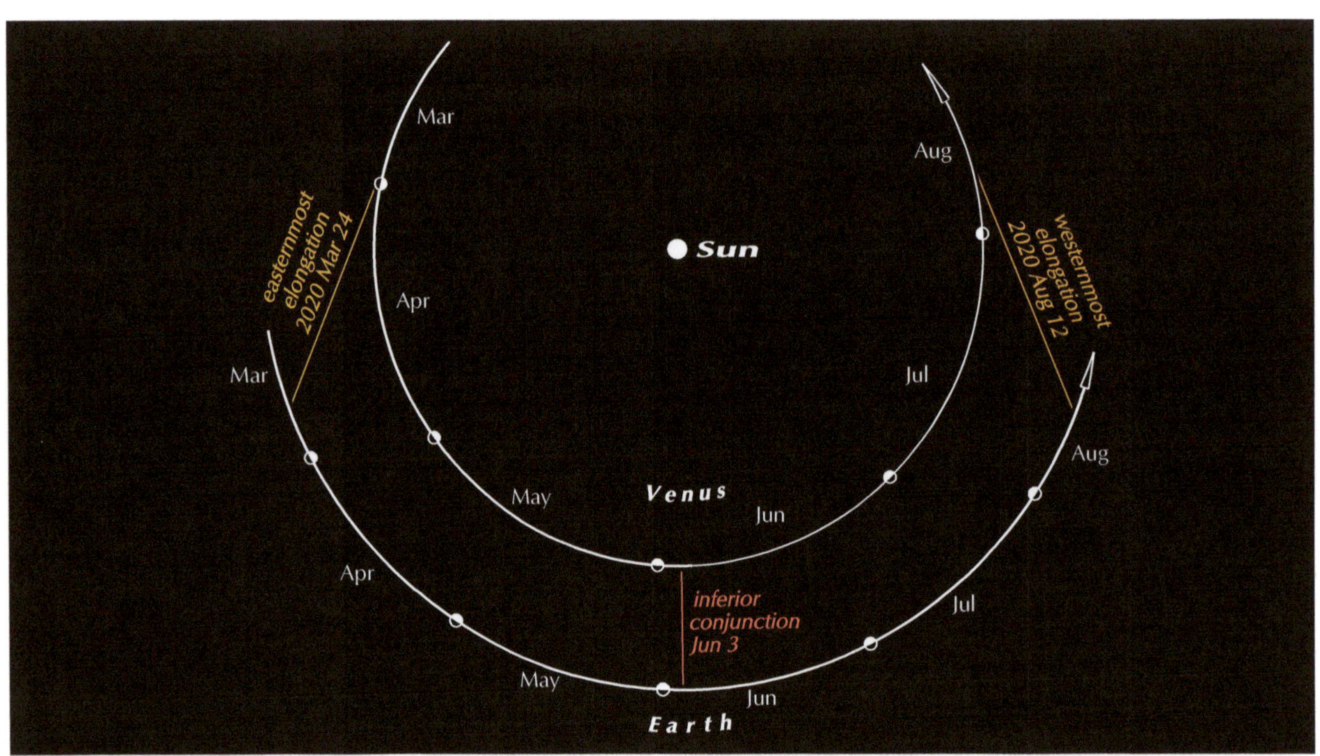

Earth and Venus from March to August 2020. The sightlines from Earth to Venus at the dates of greatest elongation are tangent to Venus's orbit. The planets are exaggerated 500 times in size, the Sun 5.

Several kinds of climax

Greatest elongation, the time when an inferior planet (Venus or Mercury) appears farthest out from the Sun's glare, is generally treated as the peak of an apparition. But there are other factors that affect observability, and so there are other moments that could be considered central to an apparition: the dates when Venus

–Is highest at dusk or dawn for your location.

–Shows the largest visible sunlit area (greatest illuminated extent).

–Is brightest.

–Is at dichotomy, appearing half sunlit.

Altitude and latitude climaxes

Great elongation from the Sun doesn't necessarily mean great altitude (height) above the horizon. It depends on the angle out along which Venus is, if you'll pardon that grammar.

This angle to the horizon changes markedly with your latitude on Earth. For north-hemisphere places, the celestial equator slopes up leftward at sunset, rightward at sunrise. For the southern hemisphere, it does the opposite. The ecliptic does about the same, but its angle varies from the equator's with the season of the year. For instance, in northern spring sunsets it is almost vertical, and in autumn much flatter. Since Venus keeps roughly to the ecliptic, its sallies out from the Sun are at greatly varying angles for all these times and places.

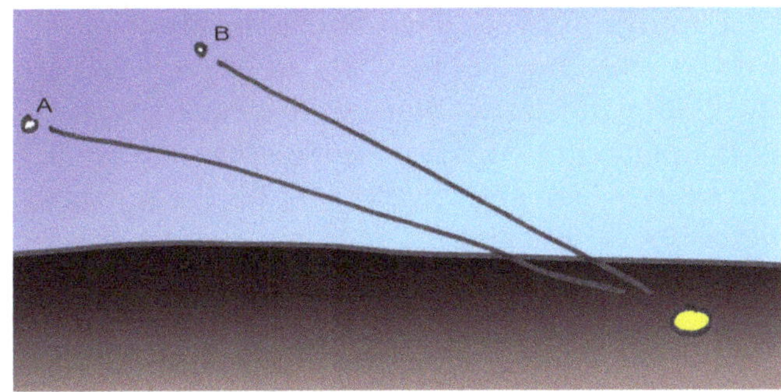

The effect of angle. A is at greater elongation from the Sun than B, yet lower over this horizon.

Here is Venus as seen on a date of greatest evening elongation from latitudes 40° north and 40° south. Very favorable, on this occasion, for Europe and North America, very unfavorable for Australasia and South Africa.

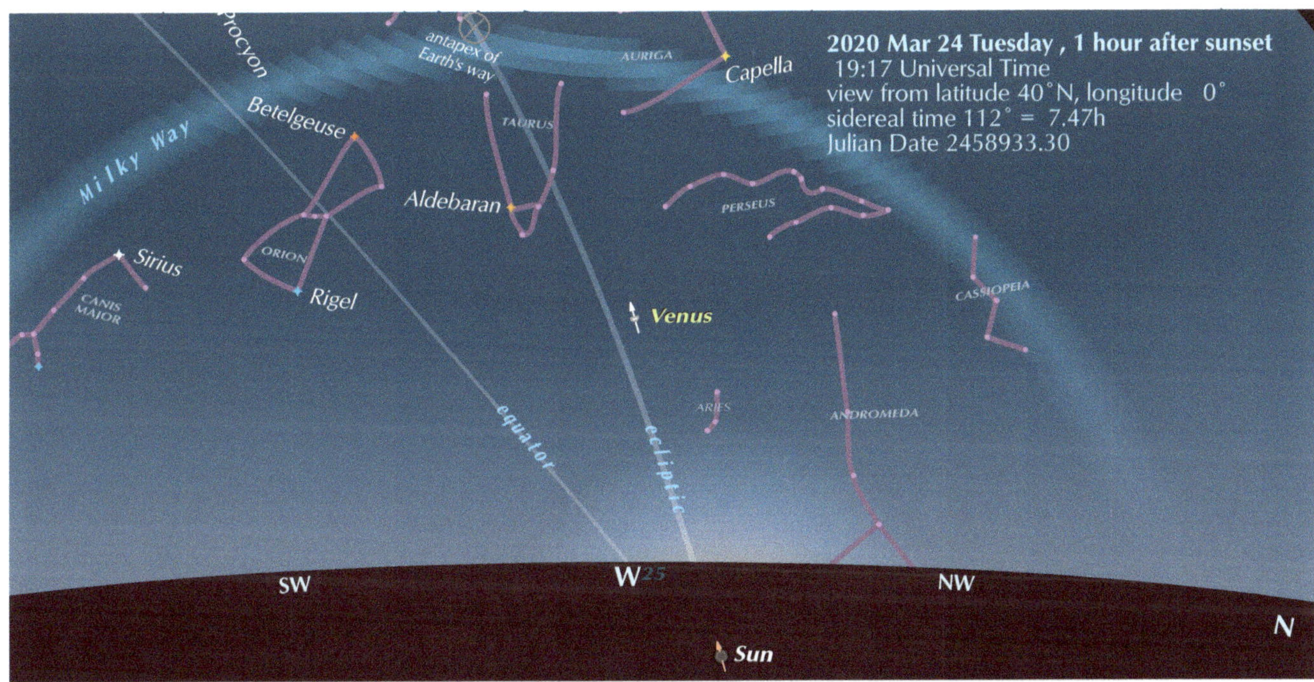

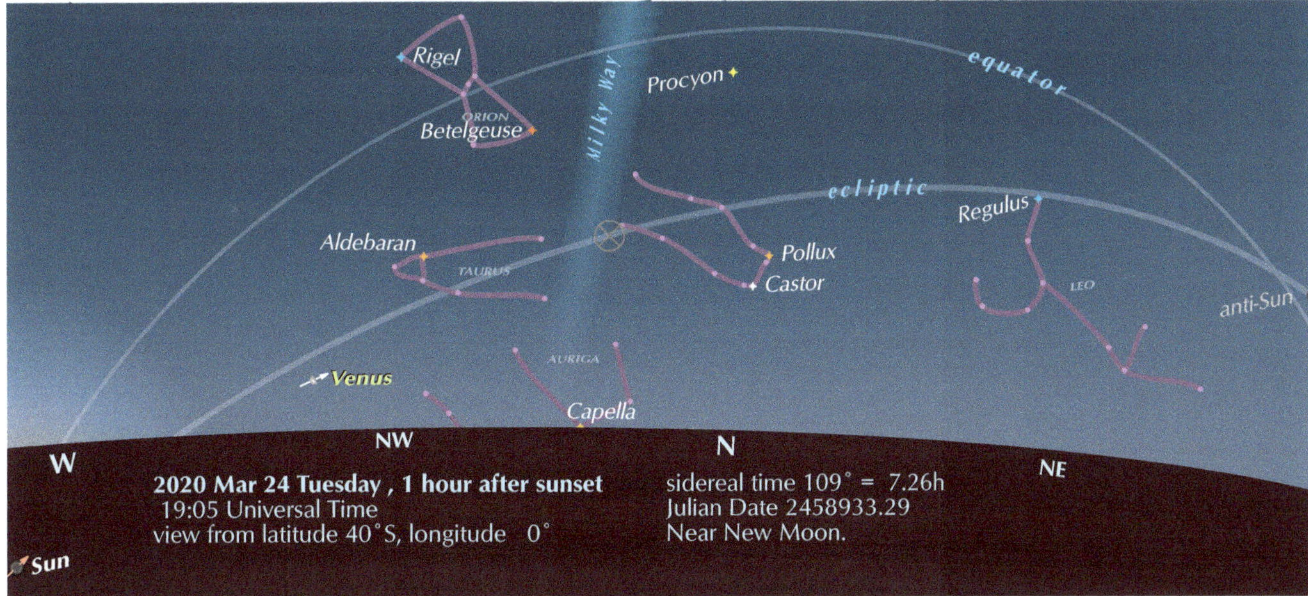

Latitude in relation to the ecliptic also makes a difference, though lesser. It can be seen that at the 2020 easternmost elongation Venus, because of happening to be in a part of its orbit north of the ecliptic, is slightly higher still for the northern hemisphere, and appreciably lower for the southern.

Even given all this, Venus at the moment of greatest elongation may not be as high above the sunset horizon as it was some days earlier or will be a few days later. For instance, in the 2026 evening sky it travels out ahead of the Sun down the most steeply south-sloping part of the ecliptic in Leo and Virgo, so that by the time of easternmost elongation it is much lower to the horizon (for the north hemisphere).

evening, latitude 40° north

2026 Jan 6 – 2026 Oct 24

Greatest illuminated extent

This means the angular area of the visible sunlit part of the planet. The time when it is at maximum may be of more interest for telescopic observers than the date of greatest elongation. This suggestion was made by Mark Gingrich, in *The Strolling Astronomer*, Jan. 2000, and *Sky & Telescope*, May 2001. So, since then, it has come to be listed in almanacs.

The planet's illuminated extent can be calculated from the distances between Earth, planet, and Sun. Its maximum happens 33 to 36 days after greatest eastern elongation or before greatest western elongation.

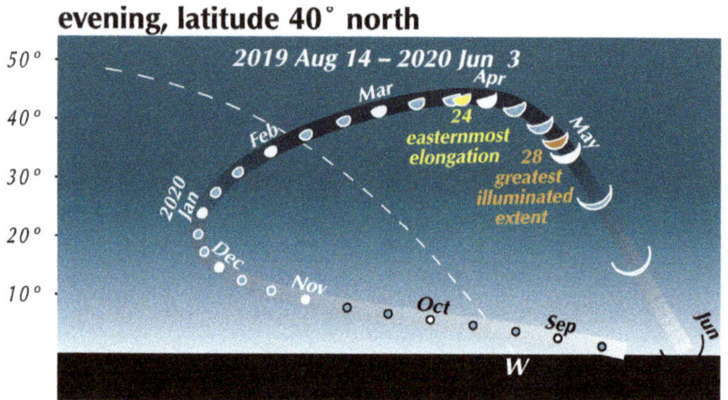

Some of the world's love stories, not all happy:

From India: Sita and Rama. His wife is stolen, he leads an army across the water (on a bridge of monkeys) to recapture her.

And from India: Rup Mati and Bâz Bahâdur. The Rajput lady is more faithful than the kinglet who carries her off.

From China: the cowherd (Altair) and the weaver girl (Vega), allowed to meet once a year across the Milky Way on a bridge of magpies.

From the Middle East: Leyla and Majnoon. Her father forbids marriage, he wanders in the wilderness mad with longing.

From mediaeval France: Héloise and Abelard. The star professor and the young flower fall in love, both suffer punishment.

Brightest

Nor is Venus at its brightest when it reaches greatest elongation. The peak of brightness comes, unsurprisingly, close to that of greatest illuminated extent, after it in the evening sky, before it in the morning sky, by between half a day and two days.

Magnitude (the astronomical measure of brightness) depends on distance from us and the Sun, and the phase angle (the Earth-planet-Sun angle), and on the albedo, but also in Venus's case on the way its cloudy surface grades toward the edges.

If we draw a graph of the magnitude of Venus, we find that on the distant side of the orbit it dims, in general, as expected—yet there is a gentle hump, a rise of a small fraction of a magnitude, during the six months around superior conjunction, when Venus is most distant and its disk smallest. Why? It's because of that phase angle. Like the Moon when Full, Venus brightens when its face turned toward us is most fully illuminated and has a minimum of shadowed area.

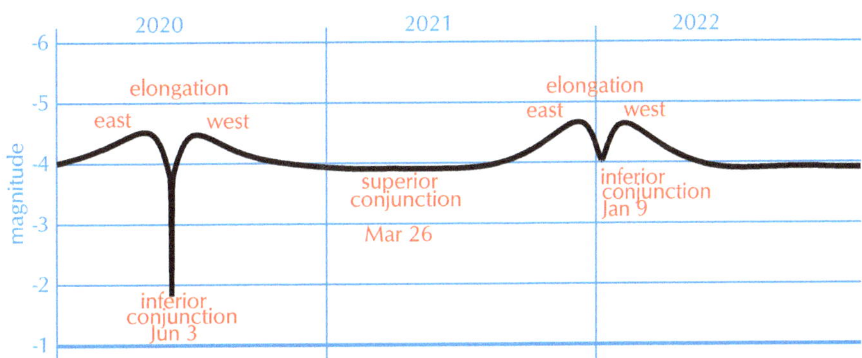

The magnitude (brightness) of Venus over several years. At inferior conjunction, the side of Venus turned toward us does not become completely black except at a transit, as in 2012; in 2020 it misses the Sun by a narrower margin than in 2022, so an extremely slender crescent is sunlit (though essentially unobservable). On the near side of its orbit, Venus moves rapidly past us, so the peaks of brightness come relatively near before and after inferior conjunction; they are about a month after easternmost and a month before westernmost elongation. Around superior conjunction, Venus is farthest from us—but there is a very slight brightening as the distant disk comes to be full-face on to us. For comparison, the magnitude of the brightest star, Sirius, is -1.4; of Jupiter, around -3; of the full Moon, nearly -13; of the Sun, nearly -27.

You may remember or misremember some of these ditties from past decades:

My Bonnie lies over the ocean, My Bonnie lies over the sea . . . Oh bring back my Bonnie to me
He sipped no sup and he craved no crumb, as he sighed for the love of a lady
Daisy, Daisy, give me your answer, do, I'm half crazy, all for the love of you
The prettiest girl I ever saw Was sipping so- Da through a straw
Ev'ry tim I think of Dinah water come to my eye
Love and marriage, love and marriage, go together like a horse and carriage
They called out "Change partners" and you waltzed away from me
Under ze bridges of Paris viz you, I'll make your dreams come tru
Stumblin' In
Got me on my knees, Layla
The only girl I ever loved has gone away
If you can't be with the one you love, love the one you're with

Dichotomy

Dichotomy means literally the "cutting in two." It is the stage at which the planet's disk is exactly cut into a light and a dark half. The terminator, the line bounding the sunlit side—the sunrise- and sunset-line on Venus—is straight, bisecting the disk, so that the bright half looks like a capital D.

It happens when the angle Earth-Venus-Sun is a right angle. This is close to the instant of greatest elongation, but not exactly the same, because of the slight eccentricity of Venus's orbit. The line from Earth tangent to Venus's orbit is not exactly at right angles to the line from Venus to the Sun. So the geometric dichotomy comes before or after the greatest elongation, by a few hours up to about three days.

Any spherical body slightly nearer to the Sun than we are will pass through this geometric position, and therefore show dichotomy. The Moon does so at its first and last quarter phases. But there is a mystery about it in the case of Venus. The moment when it geometrically should happen does not coincide with the moment when it is seen to happen.

The observed dichotomy comes too early in the evening apparition, and too late in the morning apparition. In other words, the sunlit surface of Venus appears slightly smaller than it geometrically should. This is called *Schröter's effect*, having been described by Johann Hieronymus Schröter (1745-1816). The cloudy surface of Venus grades to darker toward the terminator, and below a certain level, apparently, the eye takes it for part of the night side rather than of the sunlit side. Or, we should say, below an uncertain level, since it can vary with the observer, and from apparition to apparition. The line between light and dark may never appear straight, but slightly squiggly, because of irregularities in the clouds on Venus.

The difference between the dates of observed and geometric dichotomy is typically about a week. Apparent dichotomy has been recorded up to 10 days before greatest evening elongation, and up to 6 days after greatest morning elongation. It's interesting to record the date when you think you see it. The difference remains a slight mystery, and more data may contribute to scientists' solution of it.

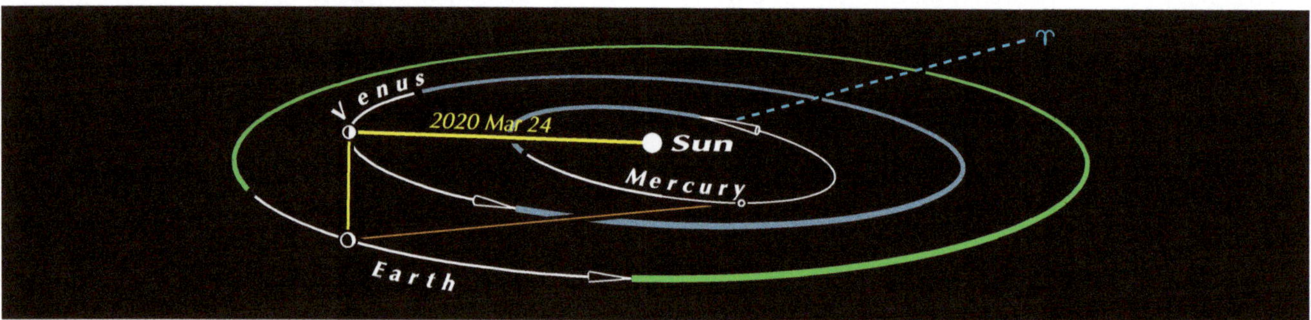

Spatial view (from 15° north of the ecliptic plane) of the planets' trajectories in March and April of 2020. The sightline from Earth to Venus, and the line from Venus to the Sun, are at March 24, the date of easternmost elongation. (The longitude of the viewpoint is chosen so that the sightline is directly away from us.) The sightline is tangent to Venus's orbit, but does not make a perfect right angle with the Venus-Sun line. But on March 27, Venus will be at dichotomy: the lines from Earth to Venus and Venus to Sun will make a right angle, but the Earth-Venus line will not be quite tangent to the Venus orbit. It happens that on March 24 Mercury, too, is at an elongation maximum (at 22 U.T.), though a western one, so we also show the sightline to it. The dashed line is the vernal equinox direction.

She was as fair as a rose at the dawning
But somehow fate hadn't meant her for me
And though I sailed with the tide in the morning
Still his heart's on the Isle of Capri

—Song "`T was on the Isle of Capri," 1934, with tango rhythm and British-English rhyming.

The retrograde episode

After easternmost elongation, Venus continues to progress eastward on the map of the sky, but that progress continues to slow, so that the Sun, also moving eastward but at its steady rate, is catching up.

Like many things in astronomy, that can be expressed in several ways. The elongation is decreasing. Venus begins to appear lower each day in the evening sky, and to set a shorter time after the Sun.

There comes the point when the eastward motion ceases. This is called the "stationary" event, though Venus is not entirely stationary on the map: it is usually moving northward or southward, at least slightly. After this, Venus moves back westward.

Then after moving over into the morning sky, Venus similarly reaches the second stationary point, at which westward movement ceases; after which it resumes eastward movement, at first slow, to catch up with the Sun, and falls into the predawn twilight.

Eastward movement is called "prograde," or "direct," since it is typical of the solar system: all the planets revolve eastward around the Sun, and most of them also rotate eastward. So westward movement is "retrograde." The time from the first stationary moment to the second is the retrograde period, and the track the planet makes on the sky in that time is its retrograde path.

All the planets, as seen from Earth, make these apparent retrograde movements. The outer planets, such as Mars, do so as Earth overtakes them on the inside. The two inner planets, Venus and Mercury, make them as they overtake Earth.

The retrograde path can be nearly straight, overlying parts of the direct path; or it can be shaped like an S or Z if the planet happens to be moving south or north across the ecliptic; or it can be a loop (crossing over itself) if the planet is passing well south or north of the Sun. Venus, because of its nearness, best shows this variety, including extravagant loops. The distant planets make flattened retrograde paths, scarcely differing from straight lines.

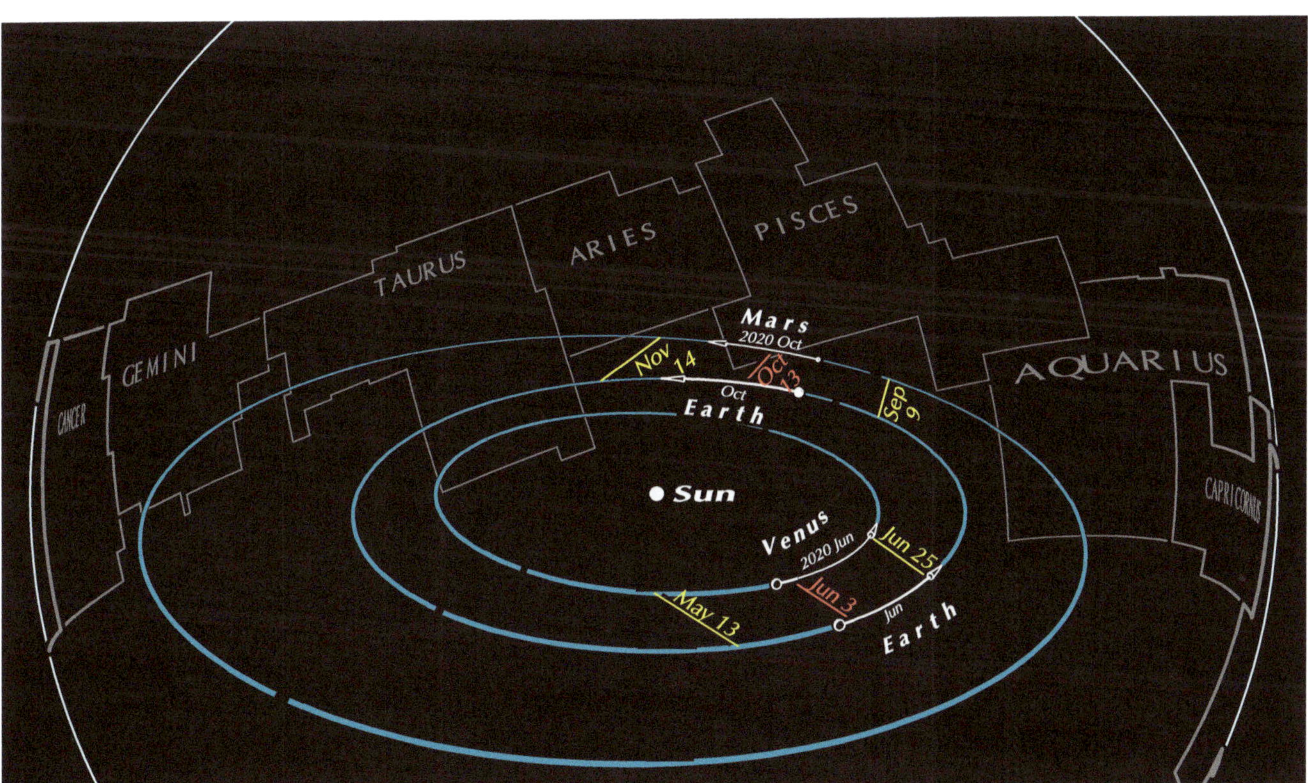

Planets appear to retrograde (move back westward) when they are overtaking Earth on the inside, or Earth is overtaking them on the inside. This spatial view shows the paths of Earth and Venus in June 2020, and Earth and Mars in October 2020. The red sightline from Earth are to Venus at its inferior conjunction of 2020 June 3, and to Mars at its opposition of 2020 Oct. 13. The yellow sightlines are from Earth to the planets at their stationary moments, the beginning and end of the apparent retrograde motions. For Venus the performance takes place, this time, against the background of Taurus; for Mars, Pisces.

The viewpoint is 6 astronomical units from the Sun and 25° north of the ecliptic plane. The boundaries of the zodiacal constellations are drawn on an imaginary sphere 2 a.u. out from the Sun.

Stationary moments

These, marking the start and end of each retrograde motion, are the instants when a planet ceases to move eastward, or westward, on the map of the sky, or in relation to the stars. But by "moving eastward or westward" we can mean at least two different things:

–Motion in right ascension: measured in the equatorial frame of reference, along or parallel to the celestial equator

–Motion in ecliptic longitude: measured in the ecliptic frame of reference, along or parallel to the ecliptic.

Because of the 23.4° angle between equator and ecliptic, the dates when Venus becomes stationary in these two senses can differ by several days.

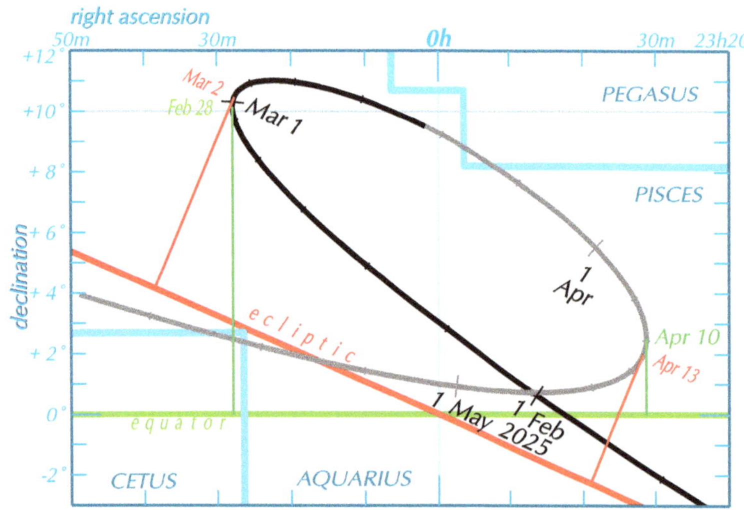

The retrograde motion of Venus in 2025. The path is drawn black when Venus is in the evening sky, gray in the morning sky, so the transition from black to gray is at the inferior conjunction. On this occasion Venus passes about 8° north of the Sun, so the retrograde loop is high, with smooth curves on either side. The green lines tangent to the curves and perpendicular to the celestial equator are from Venus at its stationary instants in right ascension, when it ceases to move eastward or westward in that frame of reference. The red lines are similarly perpendicular to the ecliptic, at the stationary instants in ecliptic longitude. The first equatorial stationary moment (Feb. 28) comes about 45 hours before the stationary moment in ecliptic longitude. The second (Apr. 10) is 58 hours before its ecliptic counterpart. These differences are reflected in the angles between the green and red lines.

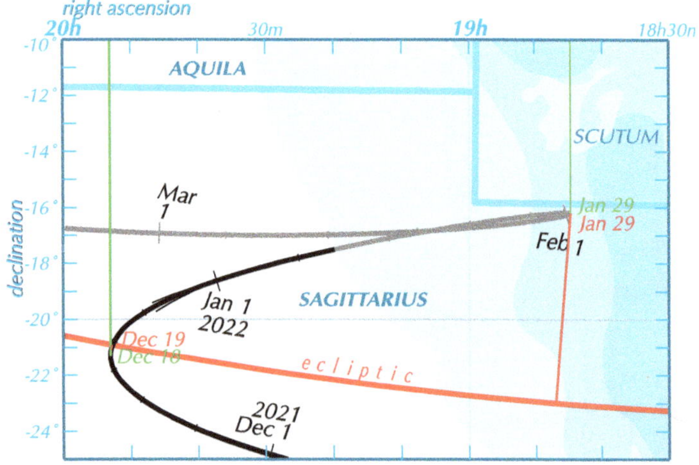

By contrast, in 2021-2022 the stationary moments of the two kinds are close together, and the lines are not much angled apart. The evening-side curve is open because Venus is ascending northward, but the eastern extreme of the curve is centered almost exactly on the ecliptic. So the red perpendicular line to the ecliptic is almost too short to see. Its angle is not much different from that of the green line, because this southerly part of the ecliptic is almost parallel to the celestial equator. The stationary moment in right ascension comes almost exactly a day before that in longitude. The morning-side end of the retrograde motion is sharp, because Venus is ceasing to move northward. The stationary moment in right ascension comes only about 47 minutes before that in longitude. The green lines perpendicular to the celestial equator go off the top of the map because Venus at this time is in the southerly part of its orbit.

Frames of reference

Positions in any map have to be measured from an "origin"—a base or zero point. For the sky, it is the vernal equinox point (in Pisces), where the Sun appears at the March equinox.

The two main frames of reference, equatorial and ecliptic, both use that origin. In the equatorial system (parallel to the spinning of the Earth), measurements are from it in right ascension and declination. In the ecliptic system (parallel to the plane of Earth's orbit), they are measured in ecliptic longitude and latitude.

In modern astronomy, the equatorial system is the default. The ecliptic system was favored in earlier centuries, and is still customary for astrology.

The equatorial suits the view in telescopes, mounted so as to stay fixed in declination. Stars drift across the field from east to west. If, say, Venus is to have a conjunction in right ascension with a star, it will be seen vertically above or below the star at that instant,

The ecliptic system is still rather appropriate for the planets, because they move in planes close to Earth's. A conjunction of planets in ecliptic longitude is likely to be nearer in time to the appulse—the instant of closest approach.

Since a planet's stationary moments in right ascension and in longitude are often several days apart, there is the question of which of them to quote. I prefer the events in longitude, since they better correspond to the ends of the planet's retrograde path.

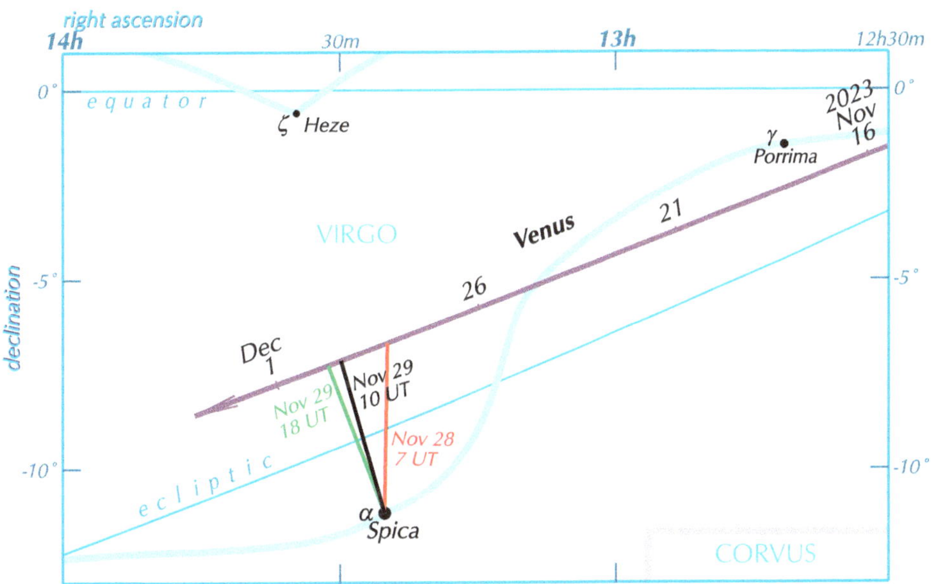

Venus passing Spica, the alpha star of Virgo, in November 2023. The red line connects them at the conjunction in right ascension; it is perpendicular to the celestial equator. The black line connects them at the appulse, the moment of shortest angular distance between them, The green line connects them at the conjunction in longitude; it is perpendicular to the ecliptic.

The westron wind is cold tonight
With a few small drops of rain.
Christ, that my love were in my arms
And I in my bed again!

The oldest recorded version of this song, from 1530, was:

Westron wynde, when wilt thou blow,
The small raine down can raine.
Cryst, if my love were in my armes
And I in my bedde again!

It may go back centuries earlier. Its tune was later used by several composers of church masses.

The swoop between Earth and Sun

The crawl around behind the Sun, from western to eastern elongation, takes 442.5 days on average; from eastern back to western it's only 141.5. That makes sense: on the far side, Venus moves with apparent slowness; on the near side it rushes past us. On the near side, not only is it larger but its time in the Sun's glare is shorter.

As it rushes at increasing apparent speed on its curve inward to overtake us, it becomes rapidly lower, in the brightest part of the evening sky.

But its shape becomes larger–and thinner. More of the dark side is turned toward us, so the remaining sunlit area becomes narrower.

There is a balance between brightness (increasing till about half way between greatest elongation and inferior conjunction), apparent size (increasing all the time), and altitude above the evening horizon (decreasing all the time). Altitude itself is a balance (as with any kind of observation not far from the Sun). The sooner after sunset, the higher the planet, but also the brighter the sky it is in. There is always the factor of sky conditions!

The ever longrt and more slender crescents, relatively huge for a planet, are dramatic in telescope or binoculars. The naked eye may be able to see Venus as not just a star-like point but slightly elongated, or even curved. This feat may need acute vision.

If you spot Venus one evening, it's a good idea to memorize its position in relation to the horizon, to help in locating it next evening when it is more difficult.

Fred Schaaf mentioned (in my *Astronomical Calendar* for 2009) a technique he used in 2001 to see the crescent with the naked eye: "center a pinhole (try sizes between 1mm and 2mm wide) in front of an eye to eliminate the rays of Venus caused by distortions in the eye's lens toward its edges."

After Venus emerges into the morning sky and goes curving away ahead of us, the swoop is played in reverse. The crescent becomes rapidly smaller, less slender, brighter, and higher in relation to the eastern sunrise horizon.

Maybe the naked-eye feat should be attempted with the rested eye of morning! The disadvantage might be not having the help of previous mornings' observations.

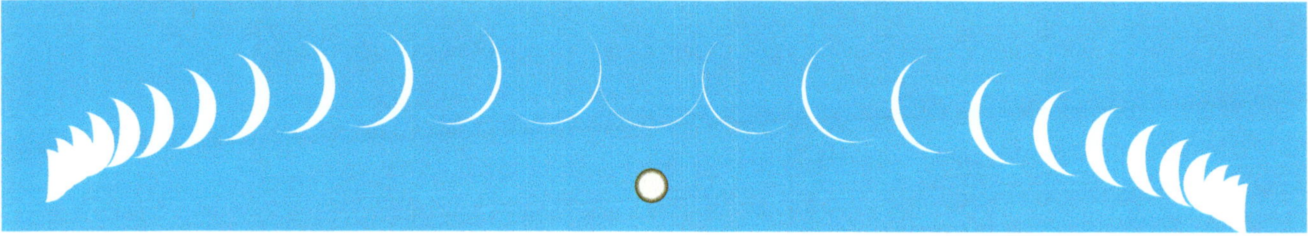

The visible part of the disk of Venus, at 5-day intervals as it passes from east of the Sun (left) to west. It is exaggerated 500 times in size, the Sun 5. The calculation is for 2025, when Venus passes 8° north of the Sun; and the positions are in ecliptic longitude–if they were in right ascension, the picture would slope down from left to right and take more space.

Hurdling the Sun

At inferior conjunction, Venus usually does not get in front of the Sun but passes north or south of it–depending on its latitude at the time, which depends on how far it is from an ascending or descending node. The latitude as seen from Earth is about 2½ times greater than the heliocentric latitude at the time, since Venus is nearer to Earth than to the Sun.

In years such as 2025, the geocentric latitude at inferior conjunction is more than 8° north. So, as seen from the northern hemisphere, Venus appears as high as it can above the Sun; from the southern hemisphere, as low as it can beneath the Sun.

High-north

```
2001 Mar. 30, 8.02°
2009 Mar. 27, 8.16°
2017 Mar. 25, 8.29°
2025 Mar. 23, 8.41°
2033 Mar. 20, 8.52°
2041 Mar. 18 , 8.61°
2049 Mar. 15 , 8.68°
```

Conversely, in years of the 2023 type, the latitude is almost 8° south. Seen from our southern hemisphere, Venus passes high above the Sun; for the north, it passes well below.

High-south

```
2007 Aug. 18, -7.98°
2015 Aug. 15, -7.84°
2023 Aug. 13, -7.69°
```

```
2031 Aug. 11, -7.53°
2039 Aug. 8, -7.36°
2047 Aug. 6, -7.17°
```

Here are the latitudes at the three other inferior conjunctions in the cycle:

```
Middling
2020 June 3,   0.49°
2022 Jan. 9,   4.85°
2026 Oct 24, -6.52°
```

2022 and 2026 have moderate clearances for the north and south respectively. 2020 has none—it's the type of year when Venus nearly transits the Sun.

Notice that, as the cycle evolves, the greatest southern clearances are gradually decreasing, and the greatest northern ones, already the more favorable, are increasing. Eventually each type will change its place in this ranking.

Since the 2025-type inferior conjunction happens near to the northern spring equinox (Mar. 20 in 2025), the angle of ecliptic to horizon is very favorable for seeing Venus near sunrise, very unfavorable at sunset. For many days before inferior conjunction, and some days after it, Venus rises well before the Sun, but sets within a few minutes of sunset. So it is possible to find Venus in the low dawn sky for many days before inferior conjunction, and then to continue seeing it progressively more easily as climbs out on the morning side.

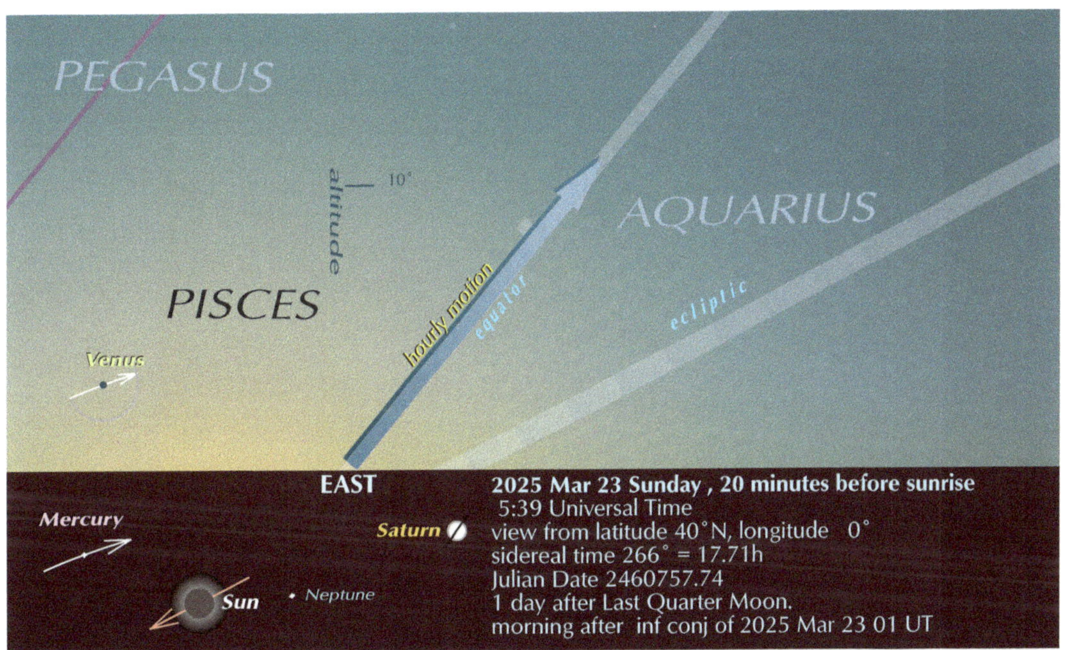

Dawn scenes, from latitude 40° north, near the 2025 inferior conjunction. The broad arrow on the celestial equator shows how much the sky seems to rotate in one hour. 40 minutes later, Venus will be higher but the Sun will be at the horizon.

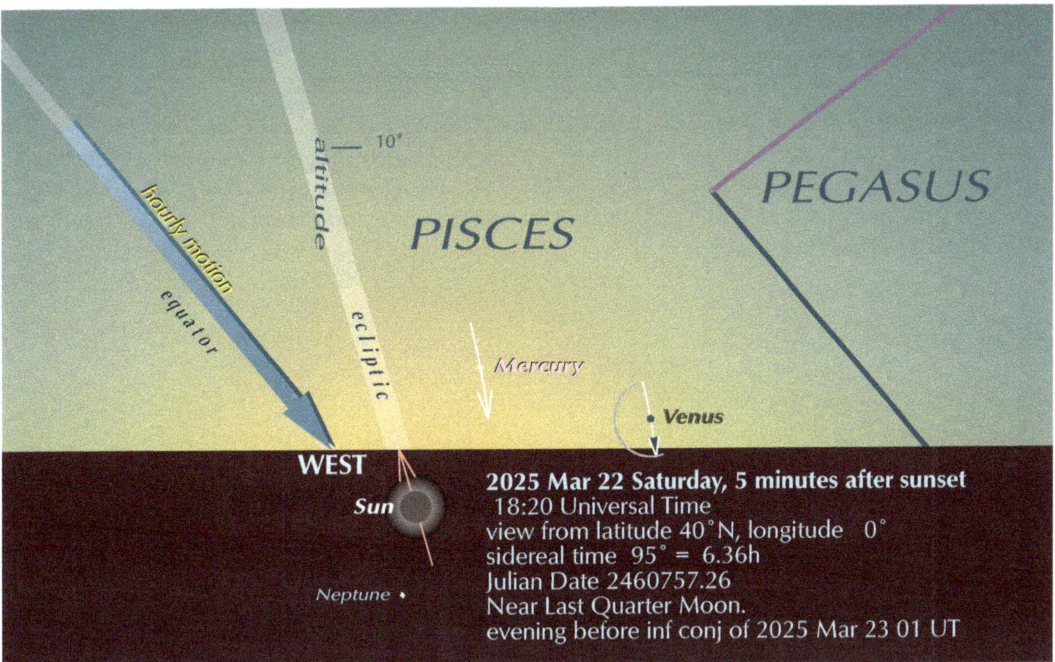

Dusk scene at almost the same date. Venus though about 8° north of the Sun is only about 2° higher than it, because of the north-hemisphere spring angle of the ecliptic; so is in too bright a sky and sets soon after the Sun.

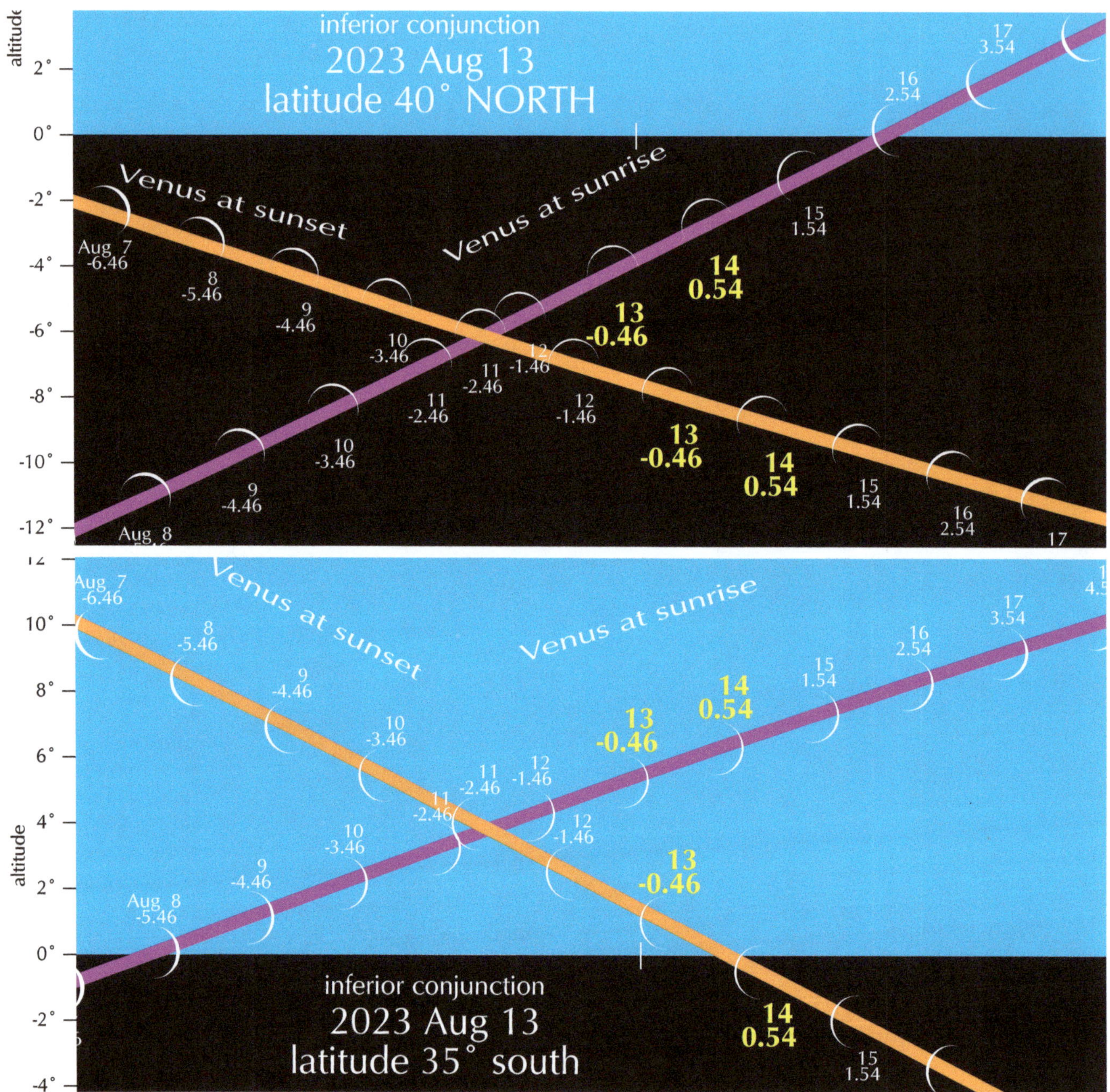

Venus's altitude at sunrises and sunsets in the days near to inferior conjunction. The planet is exaggerated 100 times in diameter; also, since the illuminated crescent is very thin around this time, it is exaggerated 4 times in width. The horizontal dimension is time from the instant of inferior conjunction; the numbers under the calendar dates are the time in days before (negative) or after that instant; those colored yellow are less than a day from it. At each date, the sunlit crescent of Venus faces toward the Sun's position on the horizon. The positions are for longitude 0, at for instance an American longitude of 90° west, Venus will be ¼ of the way back toward the previous position shown.

Seeing Venus close to inferior conjunction

Venus can be seen near or at its passage between us and the Sun. If you achieve this feat, you are seeing her tunnel from the evening to the morning sky–the Evening Star re-robing as the Morning Star.

The opportunities are when Venus clears the Sun widely: 2023-type years for the southern hemisphere, 2025-type for the northern.

Near-Sun observing has to be done with skill and caution, by figuring in advance where the tiny crescent will be in relation to the horizon, and carefully blocking the Sun with a foreground object.

To cite again Fred Schaaf's "Observers' Highlights" for March 2009, there are these possibilities:

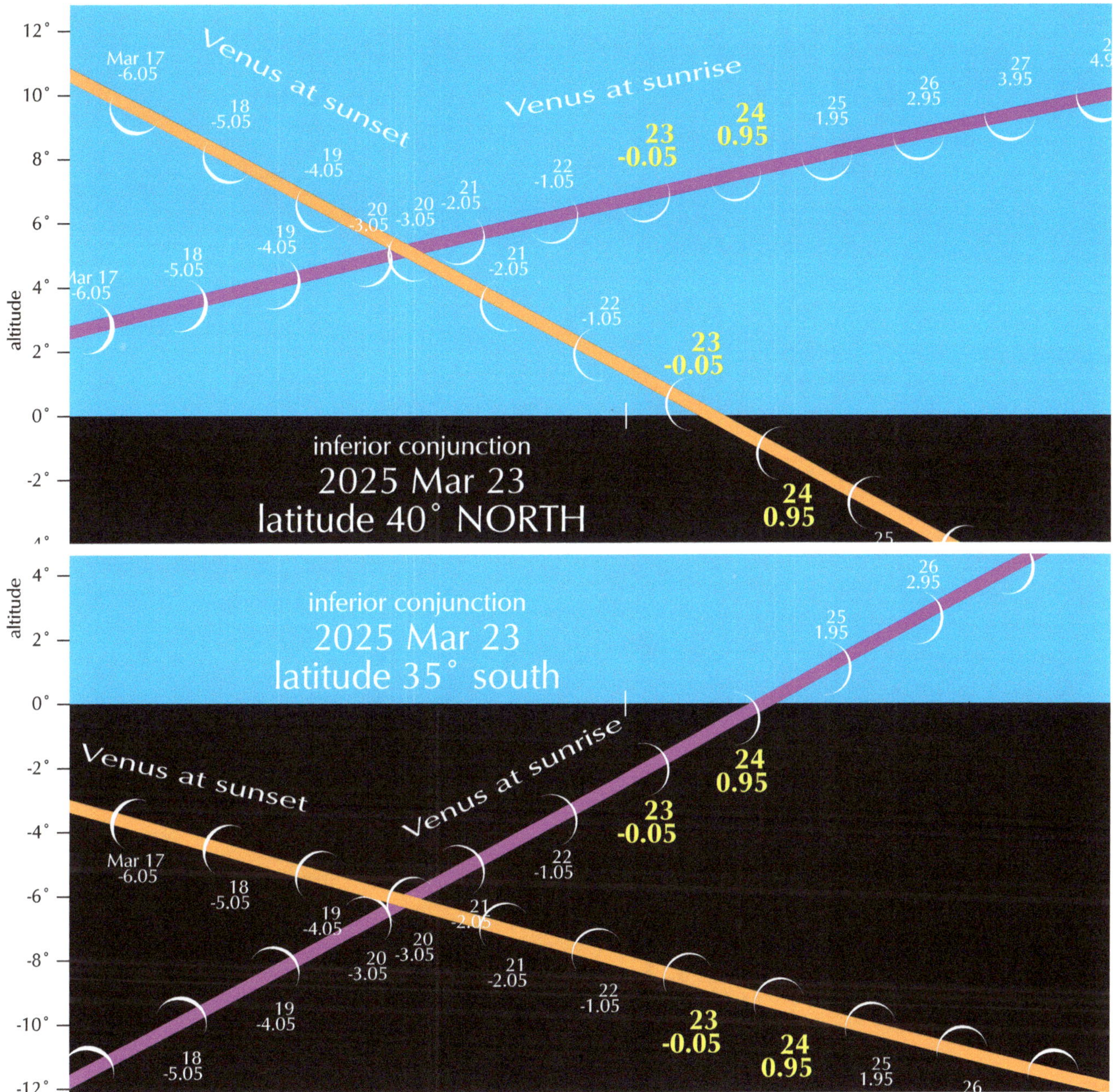

–Seeing Venus in the dawn sky before inferior conjunction.

–Seeing Venus on the morning and evening of the same day.

–Seeing Venus on either side of the same night ("something I did in 2001")..

–Seeing Venus on the day of inferior conjunction ("either naked eye or with optical aid, either with the sun below the horizon or above").

–Seeing at the exact time of inferior conjunction.

The diagrams show how in 2023 for the southern hemisphere, and 2025 for the northern, the first could be done for about a week leading to the inferior conjunction. For instance, on 2025 March 18 you could find Venus about 4° high at sunrise (lower, therefore, minutes before) and about 10° high at sunset (lower, therefore, minutes later). Over the following days, dawn altitudes would keep improving, dusk altitudes declining. The pictures for 2022 and 2026 would be like those for 2025 and 2023 somewhat flattened.

For 2020, the slopes cross at inferior conjunction. The sunset-Venuses slope is downward–that is, sunset Venuses are above the horizon before that date, sunrise Venuses after. The slopes are of different steepness; the X is tilted forward for northern hemisphere, backward for the southern. The diagrams would be identical and symmetrical if the date was at the June solstice, instead of 17 days before it.

Venus at three successive inferior conjunctions: 2017, north of the Sun; 2018, south of; 2020, almost across it.

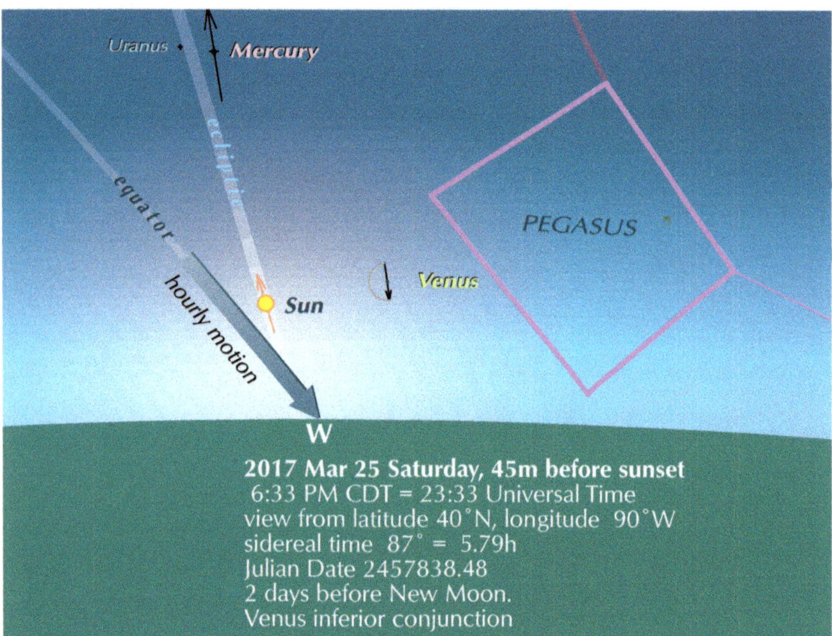

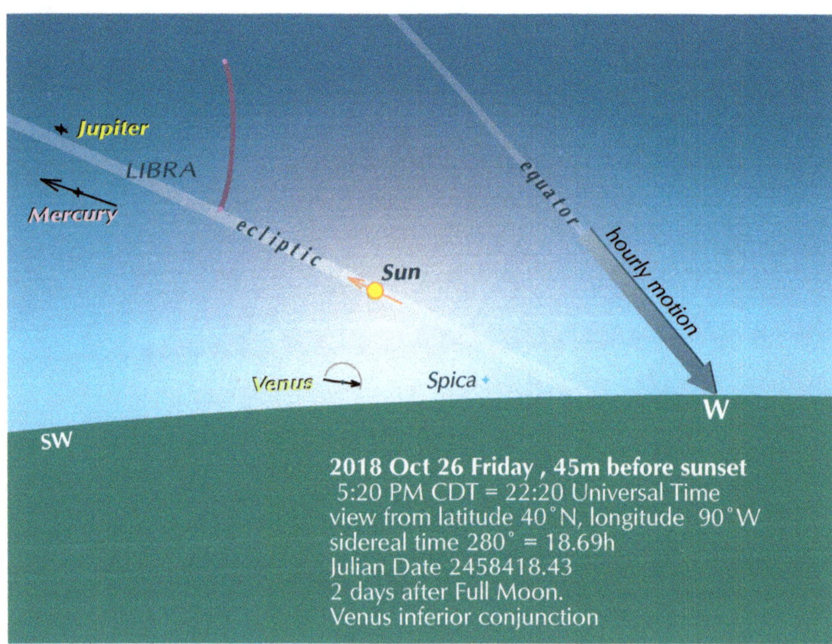

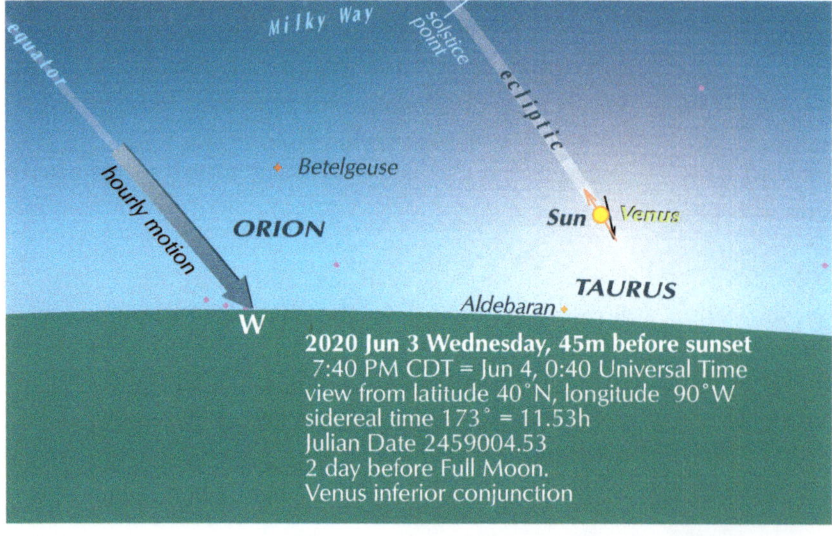

Transits

When we say that Venus, at its inferior and superior conjunctions, passes in front of and behind the Sun, we don't necessarily mean that it gets exactly in front or behind. It usually passes to the north or south—up to 8° north or south if the Sun-conjunction event happens to be half way between the ascending and descending nodes.

If, however, superior conjunction closely enough coincides with ascending or descending node, then Venus does indeed get behind the Sun. This is an occultation of Venus by the Sun: an entirely unobservable event.

Or if inferior conjunction coincides with one of the crossings of the nodes, then Venus passes literally between us and the Sun—centrally, or less so, according to how close the passage is to the node. The tiny black dot of the almost-Earth-sized planet can be seen silhouetted against the relatively enormous disk of the Sun. This is one of the famous transits of Venus. "But this deserves a chapter to itself," in the closing words of at least one of the chapters of Don Quixote; in other words, we shall have to return to this topic.

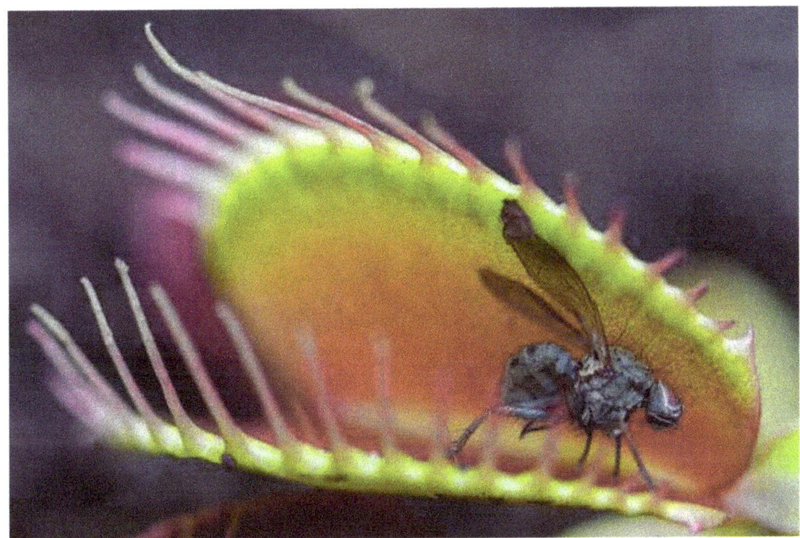

Venus flytrap. I found and painted one of these, and my article about it was printed in *South Carolina Wildlife* about 1975; but they lost my painting and used someone's photo instead.

The trapping mechanism is a modified leaf, with tiny hairs on the inner surface. At a first touch, the leaf prepares to close; if a different hair is touched within twenty seconds, the leaf snaps shut in a tenth of a second; if there are five more stimuli, it begins digesting. This ensures that energy is used only on small living prey.

It is cultivated for sale, but is now endangered in its Carolina hot wetland habitat. An older name was "tippity twitchet." The scientific name is Dione muscipula; Dione refers to Aphrodite; muscipula is Latin for a trap for a mouse, mus or fly, musca. It is the only species in genus Dione, but in its family Droseraceae carnivory has evolved in the hundreds of sundew species of genus Drosera, the sundews.

Love in a hut, with water and a crust,
Is—Love, forgive us!—cinders, ashes, dust;
Love in a palace is perhaps at last
More grievous torment than a hermit's fast

—John Keats, beginning of Part 2 of the narrative poem *Lamia*, 1820

The heart of man is capable of
Forty ridiculous kinds of love
But the heart of woman is just an ocean
Of wretched, immoderate, damp devotion

I can't recall where I heard or read this.
I certainly didn't make it up.

The eight-year cycle

There is a lovely rhythm between the orbits of the sister planets. As Earth goes around the Sun 8 times, Venus goes around 13 times, and appears, as seen from moving Earth, to go around the sky 5 times.

Let us spell that out more exactly.

–The sidereal period of Earth (the time it takes to go around the Sun) is 365.256 days. Multiplied by 8, that is about 2,922 days.

–The sidereal period of Venus is 224.701 days. Multiplied by 13, that is about 2,921.

–The synodic period of Venus (the time it takes to return to a given position in relation to the Sun, as seen from Earth) is 583.92 days. Multiplied by 5, that is about 2,920.

Those numbers coincide well, though not exactly: a cycle of about 2,921 days: close to 8 Earth-years and 13 Venus-years.

As a result, if Venus on a certain date is giving us a certain show, such as passing near Aldebaran or lying as far north as it can or shining highest above New York's sunset horizon, then 8 years later it will be doing the same.

So each year of the 8-year cycle has a character, a distinctive pattern, shaped by the timing of the planet's movements. The phenomena of 2000 repeat, very similarly, in 2008, 2016, 2024, and so on.

The coincidence is not exact. A span of 8 Earth-years is almost one day longer than 13 Venus-years. So the cycle has to evolve. The heliocentric events (such as perihelion and ascending node) arrive, in the corresponding year of the next cycle, about a day earlier. The geocentric events (such as the Sun-conjunctions and greatest elongations) become a bit more than 2 days later. They therefore occur at slightly shifted positions in the sky.

So the strong pattern of, say, the 2000-type year gradually changes, over the centuries.

There are longer and closer coincidences. 243 Earth-years and 395 Venus-years are both close to 88,757 days.

But Earth-Venus harmonies, beautiful though they are to us, are not stable. They evolve in one direction and do not swing back cyclically, so they are not locked, like the 2-3 resonance between the orbits of Pluto and Neptune. They are a feature of the present few thousand years, an eye-blink in the four-and-a-half-billion years of astronomical time.

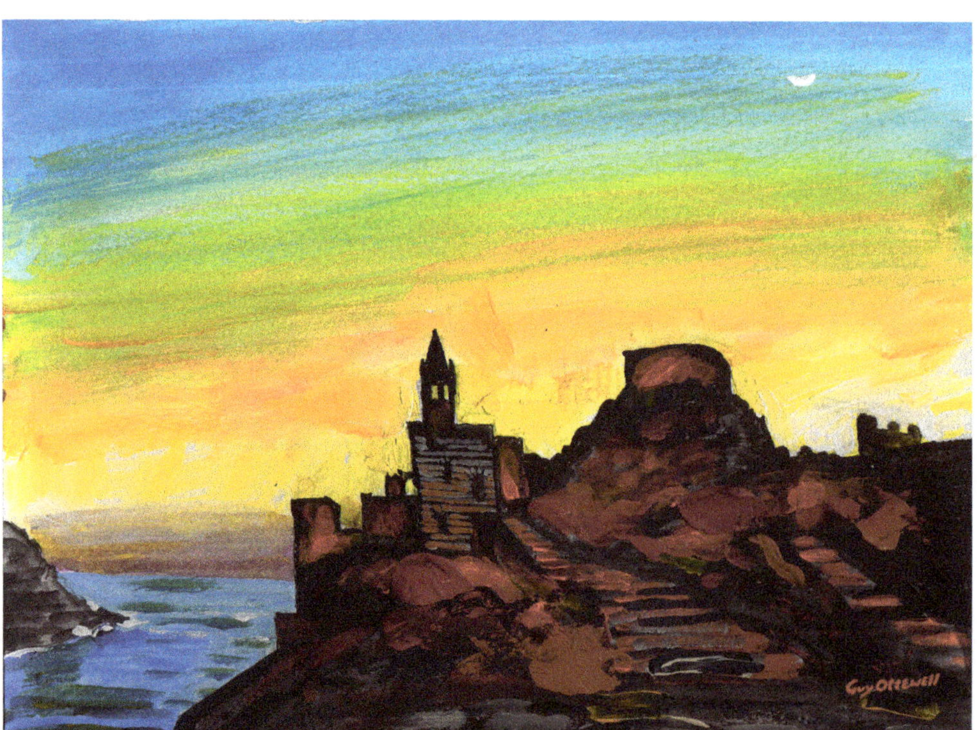

Porto Venere, on the Ligurian coast of Italy, silhouetted against the sunset of 2006 April 3. Five days earlier, from out on the Mediterranean, we saw that Moon eclipse the Sun. As to the name: one story is that there was on this promontory a temple of Venus; another is that there was a hermit, Saint Venerius. He is patron of the Gulf of Spezia and of lighthouses, and his festival is celebrated here on September 13.

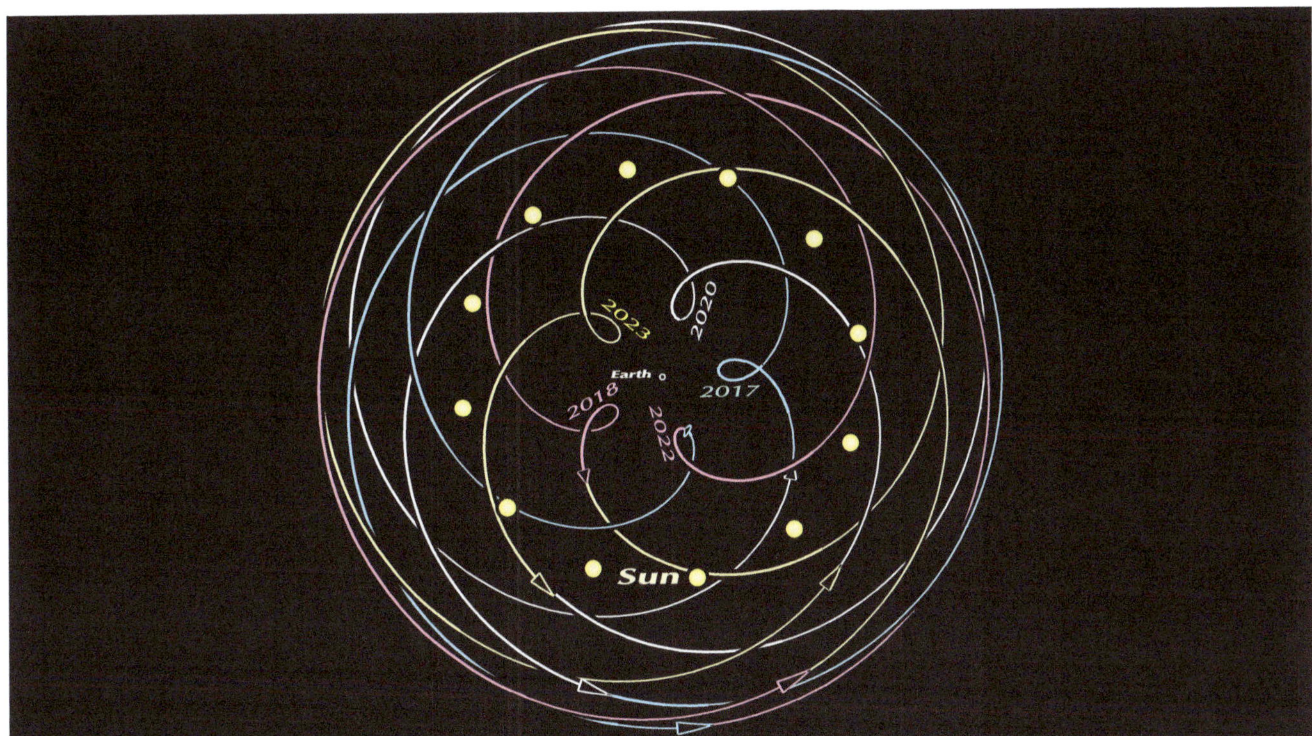

The "Pentagram of Venus." This is a geocentric space view: the Earth is held still. So the Sun appears to go around Earth, in 12 months. Venus also appears to go around Earth, 5 times in the course of 8 years, 2016 to 2023. The paths of Venus in the years are shown in different colors, starting at the lower left with 2016 in white, and ending at the same point at the end of 2023's yellow path. During the 8 years, Venus 5 times makes an Earth-ward loop, at the inferior conjunctions of 2017 Mar. 25, 2018 Oct. 26, 2020 June 3, 2022 Jan. 9, and 2023 Aug. 13. If you draw lines connecting neighboring loops (2017 to 2020 . . .), they form a pentagon. But if instead you connect them in time sequence (2017 to 2018 to 2020 . . .), they form a pentagram. This is because the synodic period of Venus is about 1.6 years so from the inferior conjunction of 2017 to that of 2018 it goes around the circle 1.6 times. Similarly the 5 superior conjunctions of the time span, when Venus is opposite to Earth across the Sun, are spaced around the outer limit of the diagram; and, if connected, would make a pentagram.

The view is from ecliptic north, and the track of Venus is drawn thinner when it is south of the ecliptic plane. This geocentric kind of picture, now unfamiliar to us, is geometrically equivalent to the heliocentric picture, though gravitationally unlikely. It is similar to the conception of the cosmos generally accepted before Copernicus, Kepler, and Galileo. You can see why, to make that system approximately work, the presumed circular orbits of the planets around Earth had to be elaborated with "epicycles," somewhat similar to the retrograde loops.

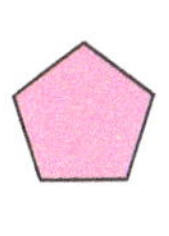
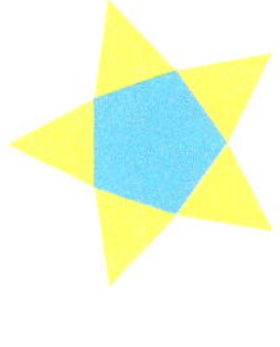
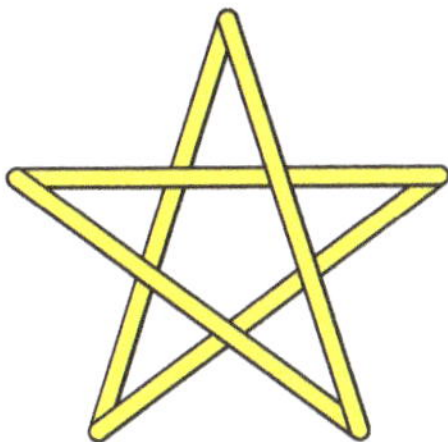
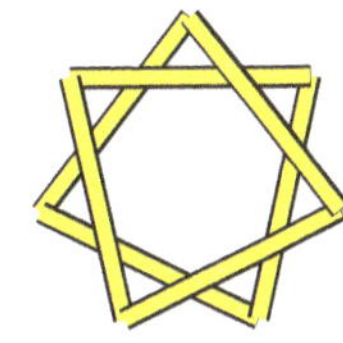
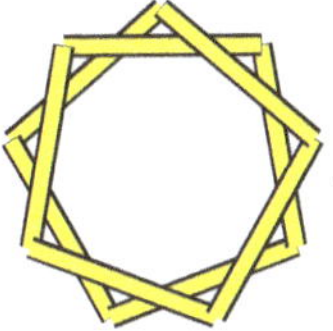
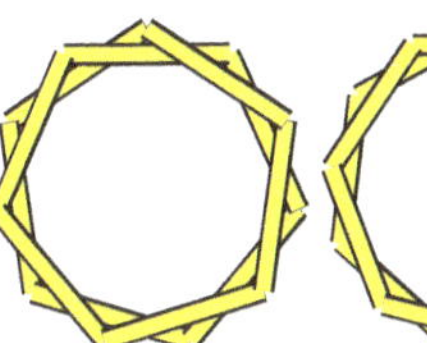
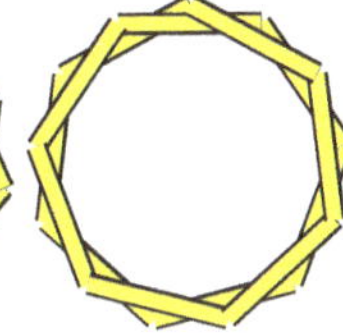

A pentagon, or five-sided polygon; a stellated pentagon, with triangles extending each side; and a pentagram, whose interior and outline are a pentagon and a stellated pentagon. It can be drawn as a continuous path, with strokes that cross over each other. Each stroke connects to the next point but one. The other shapes with odd numbers of points (shown are 7, 9, 11, and 13) can be drawn likewise. By contrast for instance the six-pointed star, or hexagram, has to be drawn by superimposing two triangles.

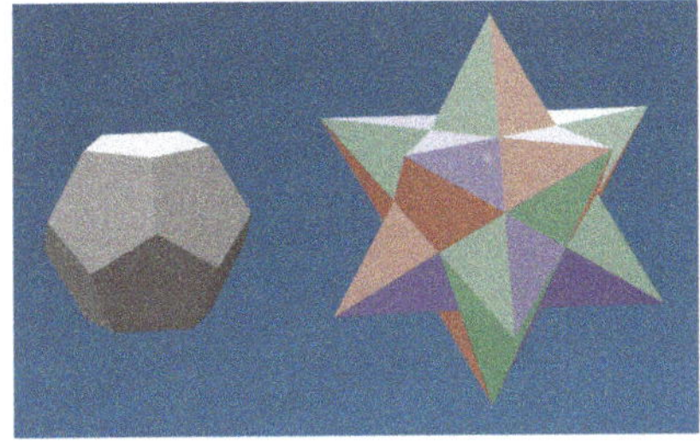

And, just for fun and because they are so pleasing to look at: a dodecahedron, 12-faced, each face a pentagon; and the same stellated. (All this about polygons and polyhedrons derives from my study of them for the *Astronomical Calendar 2015* cover picture and story about stellated stars.)

Fibonacci

The numbers 5, 8, and 13 happen to be successive numbers in the famous Fibonacci series, which goes like this:

1+1=2, 1+2=3, 2+3=5, 3+5=8, 5+8=13, 8+13=21, and so on.

The ratios of the successive numbers converge:

1 / 1 = 1
1 / 2 = 0.5
2 / 3 = 0.6666666 . . .
3 / 5 = 0.6
5 / 8 = 0.625
8 / 13 = 0.6153846 . . .
13 / 21 = 0.6190476. . . . ..
21 / 34 = 0.6176471 . . .
34 / 55 = 0.6181818 . . .
55 / 89 = 0.6179775 . . .
89 / 144 = 0.6180556 . . .
144 / 233 = 0.6180258 . . .
233 / 377 = 0.6180371 . . .
377 / 610 = 0.6180328 . . .
610 / 987 = 0.6180344 . . .
987 / 1597 = 0.6180338 . . .
1597 / 2584 = 0.6180341 . . .
2584 / 4181 = 0.6180340 . . .
4181 / 6765 = 0.6180340 . . .

What they converge on is the Golden Ratio, or Phi, one of the important "irrational" or never-ending numbers (like Pi): 0.61803 . . . The Golden Ratio is the only number that is mirrored by its own inverse: 1 / 0.61803 = 1.61803 and 1 / 1.61803 = 0.61803. If any two things have this proportion, such as sides of a rectangle, the larger of them has the same proportion to their sum, and so on *ad infinitum.*

Fibonacci numbers, and the Golden Proportion, are found in the spiral arrangement of leaves and sunflower heads and pineapple scales, the successively larger chambers of nautilus shells . . . and have been discerned in the proportions of Greek temples, the divisions of books of the Aeneid . . .

Is there a physical reason for their influence on the harmonies of Venus? There are many other such resonances between orbits in the solar system. There are longer and more accurate Earth-Venus resonances. After 243 Earth-years and 395 Venus-Years, their relative positions differ by only 0.8°.

Octave

The eight-year cycle is reminiscent of the octave, fundamental to music. In a scale of notes, as on the piano, each note produces a discord if sounded with those closest to it–the closer the neighbor, the nastier the discord–but with the note one octave higher of lower it produces a concord so perfect that they are regarded as the same note: A and upper A, C and under C . . . In the Venus cycle, each year is dissimilar to its neighbors, and then is repeated in the first year of the next cycle.

This analogy, though attractive, is not fully valid. The word "octave" is somewhat misleading. Our scales (major and minor) consist not of eight tones but of seven: then the first is repeated in the eighth, the beginning of the next octave. The octave is like a stairway of eight levels with seven risers between them. The Venus cycle is like a stairway of nine levels with eight risers between them.

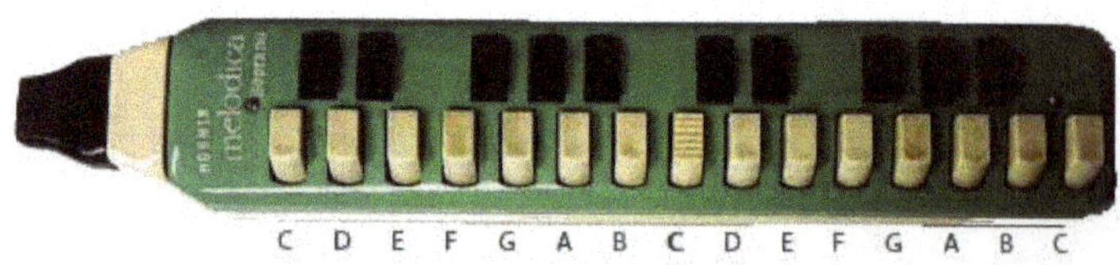

A keyboard of two octaves, on an instrument called a Hohner soprano melodica. Midd;e C. ending the lower octave and starting the upper octave, is corrugated so that the finger can feel it as a reference point. The black keys are the semitones between the notes of the major scale.

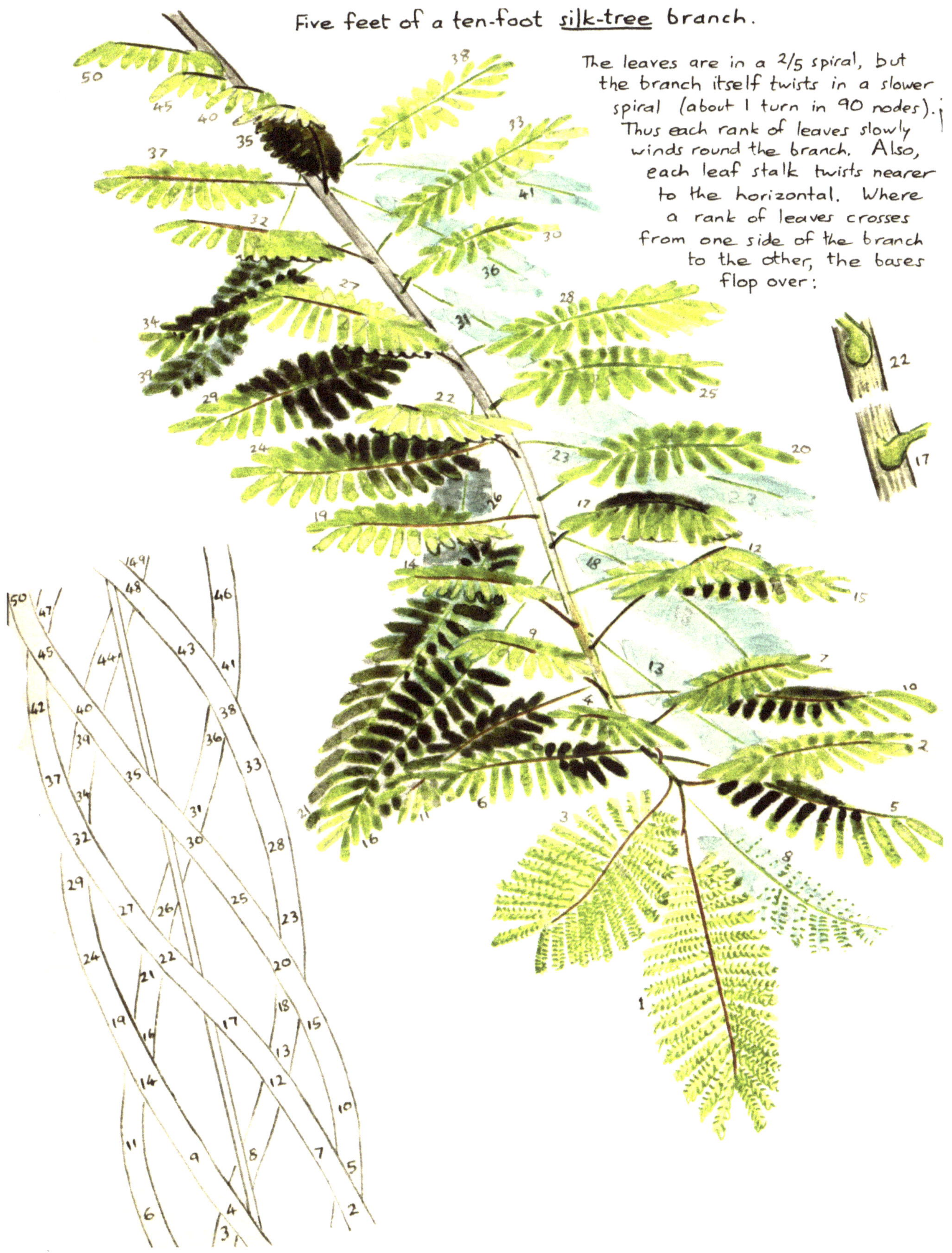

One of my many studies of plant spirals
Branch of a silktree, *Albizia julibrissin*.

Loops and the constellations they take Venus into

The constellations used to be informal areas of the sky. Since 1930 they have had official boundaries, adopted by the International Astronomical Union, so by being "in" a constellation we mean position in relation to these artificial lines.

The ecliptic goes through the twelve constellations of the zodiac: Pisces, Aries, Taurus, Gemini, Cancer, Leo, Virgo, Libra, Scorpius, Sagittarius, Capricornus, Aquarius. (That list is in eastward or "right-to-left" order, because the solar-system bodies travel generally eastward.) The ecliptic also, after a narrow bit of Scorpius, goes through a wider area defined as belonging to Ophiuchus. We can speak of the planets as traveling within a broader zodiacal band, centered on the ecliptic.

But, because of their orbital inclinations, planets and the Moon take apparent paths that can curve up and down into constellations other than those through which the ecliptic passes. Those into which they can wander are:

–On the north side of the ecliptic, Auriga, Scutum, Pegasus.

–On the south, Cetus, Orion, Hydra, Sextans, Corvus, Crater.

It's Venus that does most of this ex-zodiac venturing.

Jean Meeus, in his *More Mathematical Astronomy Morsels*, 2002, p. 333, discusses in detail the de-facto "22 zodiacal constellations," that is, the traditional 12 plus Ophiuchus plus the 9 non-ecliptic ones. He gives dates of episodes as short as 4 days when Venus was or will be in Pegasus (1798), Scutum (2014), Corvus (2058), and Crater (2122), and a small chart of the 2014 Scutum episode.

The part of Venus's track around the time of inferior conjunction is the part that most diverges from the ecliptic, and is also the part that most clearly displays the gradual evolution from cycle to cycle.

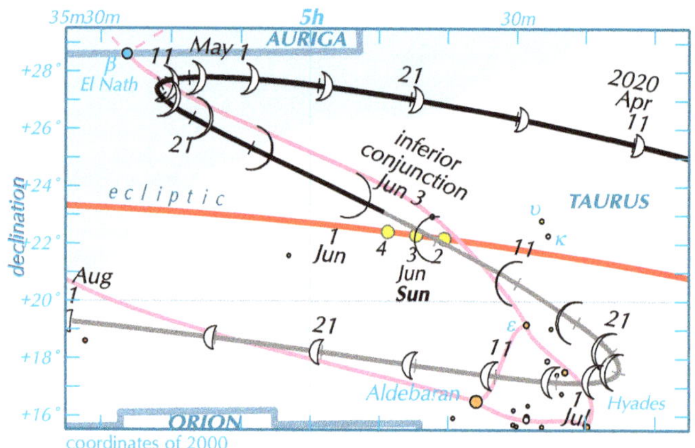

Tracks of Venus around the times of the inferior conjunctions of 2020, 2022, 2023, 2025, and 2026. The retrograde movement takes five different forms—S shape, loops north or south of the ecliptic, and transitional. The planet's sunlit crescent is exaggerated 100 times in size.

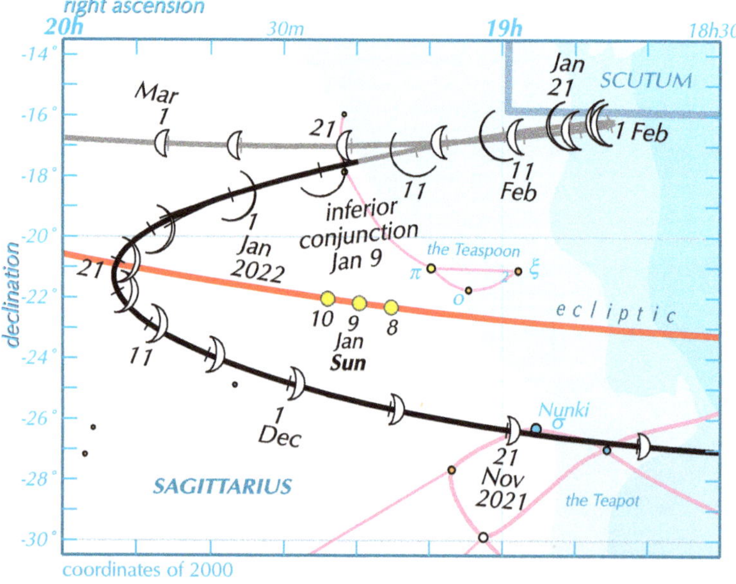

Yak Zan Bas Ast!

"One Woman Enough Is!"—a pious passenger, rebuking a bus driver who boasted that he had a wife in each town along his route. Persian *zan*, like French *femme*, means both "woman" and "wife."

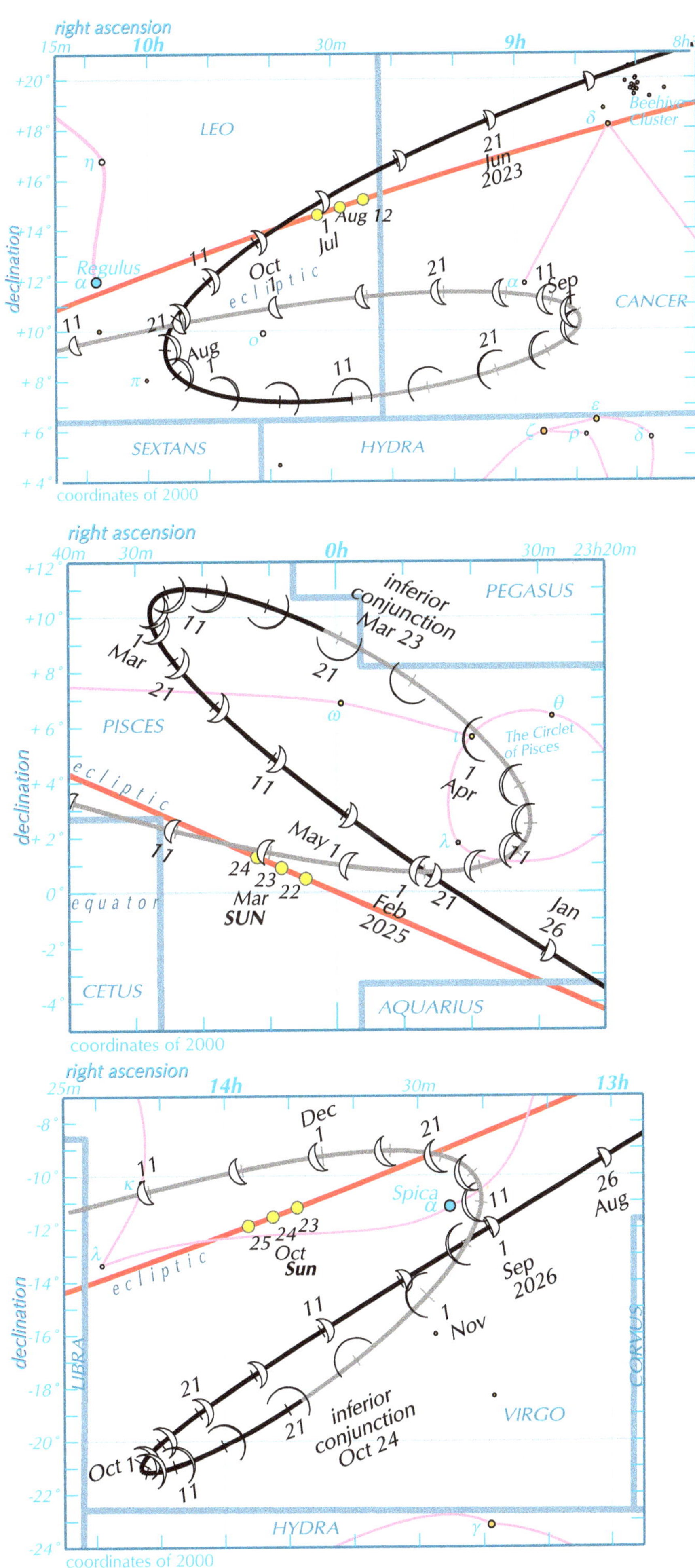
right ascension
15m
10h
30m
9h
8h30m
+20°
+18°
+16°
+14°
+12°
+10°
+8°
+6°
+4°
declination
LEO
Beehive Cluster
21 Jun 2023
Aug 12
1 Jul
Regulus
Oct 1
ecliptic
11
21
Sep
CANCER
Aug 1
SEXTANS
HYDRA
coordinates of 2000
right ascension
40m
30m
0h
30m
23h20m
+12°
+10°
+8°
+6°
+4°
+2°
0°
-2°
-4°
inferior conjunction Mar 23
PEGASUS
1 Mar
PISCES
The Circlet of Pisces
1 Apr
ecliptic
May 1
24 23 22
Mar
SUN
1 Feb 2025
Jan 26
equator
CETUS
AQUARIUS
coordinates of 2000
right ascension
25m
14h
30m
13h
-8°
-10°
-12°
-14°
-16°
-18°
-20°
-22°
-24°
declination
Dec 1
21
11
Spica
26 Aug
25 24 23
Oct
Sun
ecliptic
1 Sep 2026
1 Nov
LIBRA
CORVUS
inferior conjunction Oct 24
VIRGO
Oct 17
HYDRA
coordinates of 2000

Loops evolving

As the eight-year cycle gradually changes, events such as inferior conjunction are displaced 2 or 3 days backward in the calendar, and positions drift slowly westward on the map of the sky.

This is not easy to see in the long forward tracks of different years, which are almost parallel to the ecliptic and almost overlie each other. But it becomes obvious in the retrograde parts of the tracks, because of their extravagant departures from the ecliptic.

Pur mai non sentesi
Felice appieno
Chi su quel seno
Non liba amore!

Yet there's no happiness
Full as it ought to be
For him who fails to drink
From that bosom love!

Conclusion of the aria "La donna è mobile" in Verdi's opera *Rigoletto* (1851)

Eppur si muove.
And yet it does move.

Galileo is said to have muttered this on leaving his 1633 trial; he had been forced to recant his belief in the Copernican system in which the Earth and other planets move around the Sun.

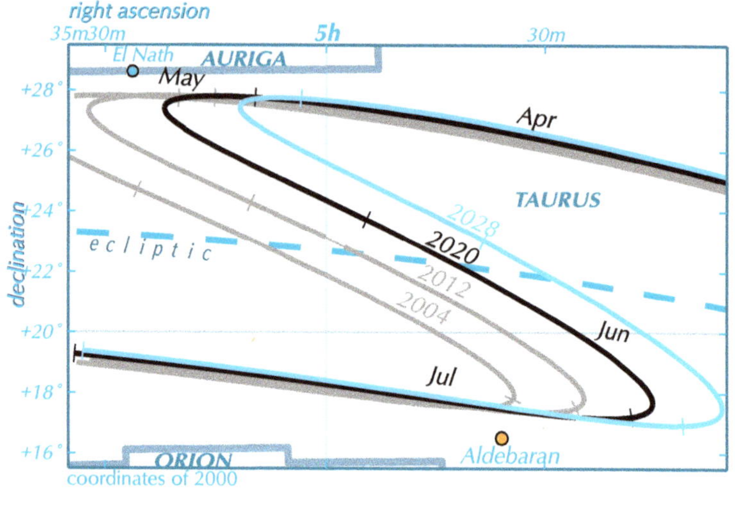

Venus's track around the time of the 2020 inferior conjunction—and at the previous two occasions of this cycle year (gray) and the next future one (blue).

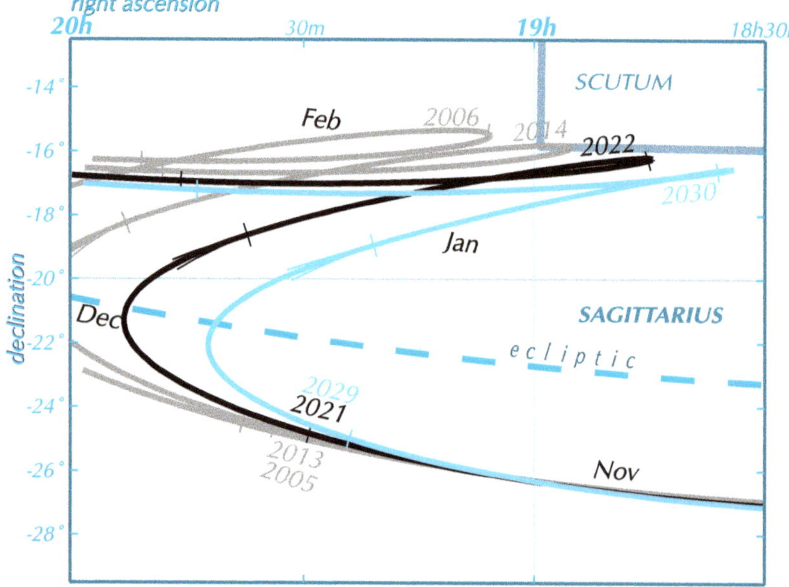

Venus was barely inside Scutum for a few days in 2014.

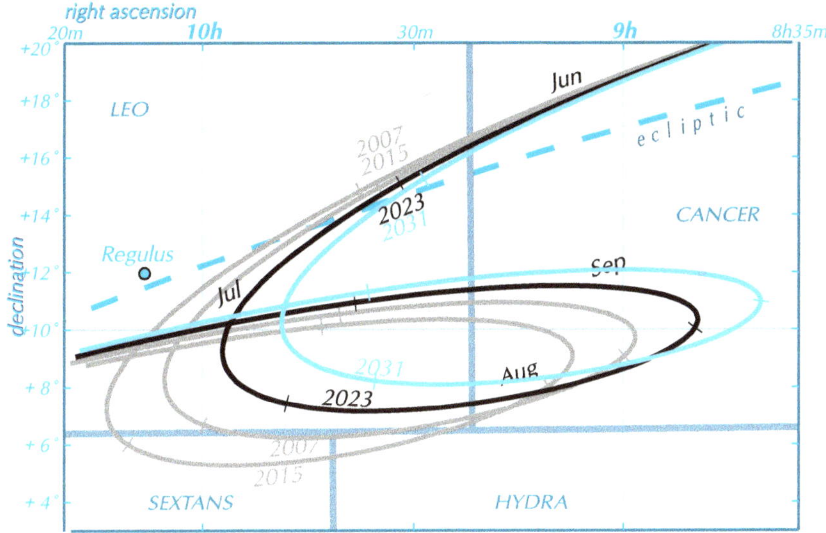

Venus spent time in Sextans and Hydra in 2015, 2007, and earlier passages, but not later ones.

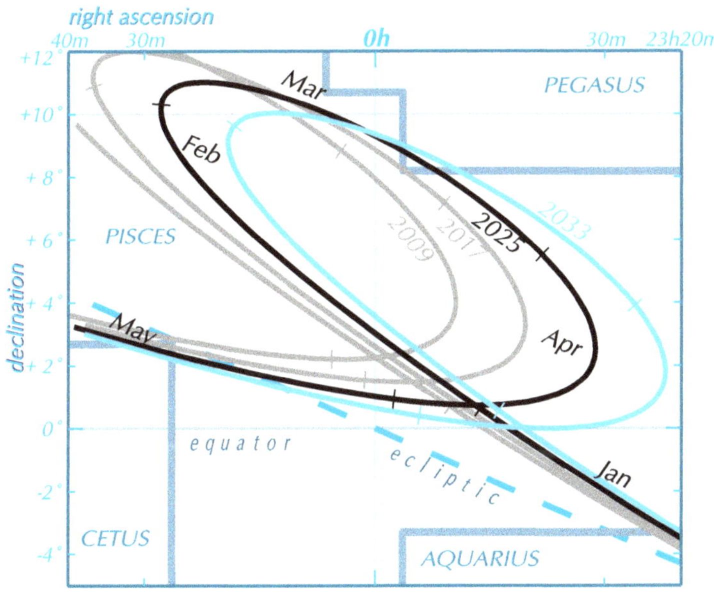

Venus barely touches Cetus and Pegasus in 2017, enters them more deeply in succeeding passages.

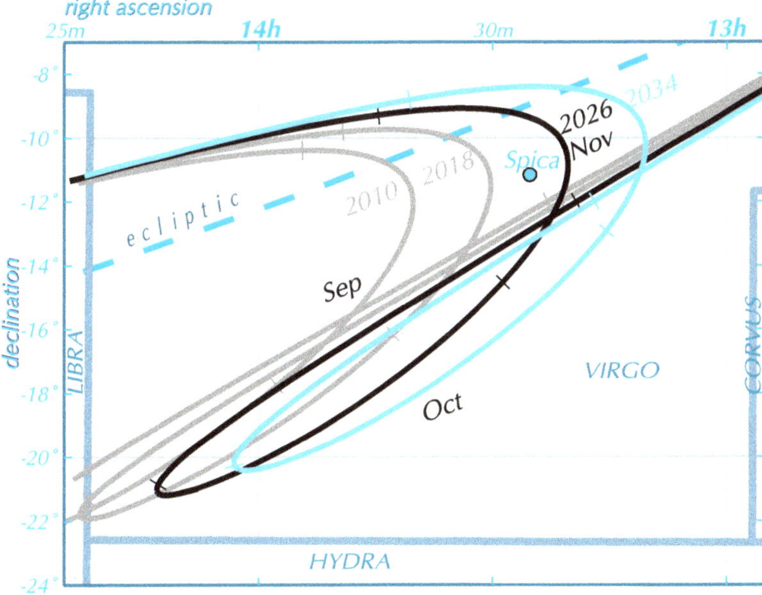

In future centuries, this south-dangling loop will sweep the corners of the small constellations on the back of the water serpent Hydra: Corvus, and then, farther west, Crater—and Sextans, which the 2023-type loop has just finished visiting.

Northness and southness

Because of the 3.4° inclination of its orbit to the ecliptic, Venus can reach that ecliptic latitude north or south, as seen from the Sun. Seen from Earth, which can come nearer to it, Venus can appear at a greater latitude north or south of the ecliptic plane: as much as 8.4°.

This has to happen around the time when Venus is nearest to us, not far from inferior conjunction. So the extremes of declination (northness and southness from the equator) also have to be not far from inferior conjunction. Fortunately, Venus can be observed at or near these northernmost and southernmost times, because the time between easternmost or westernmost elongation and inferior conjunction is relatively short.

The northernmost declination in the current 8-year cycle is 27.82°, on 2020 May 4.

That happens because Venus is not only traveling the northernmost part of the ecliptic, in Taurus, but is in a part of its orbit that is nearly highest north of the ecliptic plane (the northernmost latitude was on Apr. 10). That in turn results from its having been at the ascending node of the orbit on Feb. 15. It is another consequence of the orientation of the whole Venus apparition.

The date of northernmost latitude is not long after easternmost elongation, and then the northernmost declination is reached shortly before the stationary moment, so Venus is still well out in the evening sky and easily observable.

The southernmost declination is -27.24° on 2021 Nov. 6, in Sagittarius. That also is close to easternmost elongation, and favors southerners.

In other years, Venus's northernmost and southernmost points are nearer to the ecliptic, or on the wrong side (south of the northern part, or north of the southern part), so that the extremes of declination are not great.

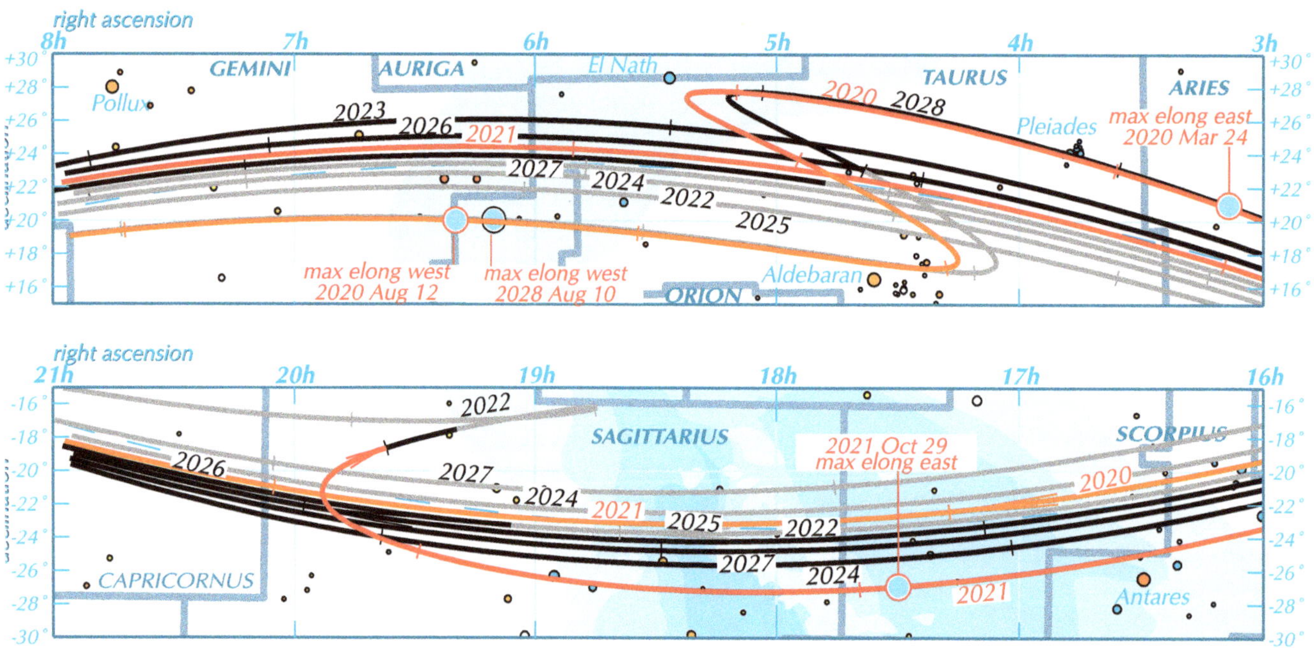

Northern and southern parts of Venus's journeys. In red are the path for 2020, when Venus reaches northernmost, and 2021, when it reaches southernmost. In black are the other 6 years of the cycle; also 2028, almost repeating 2020 and beginning the next cycle. The colors are paler when Venus is in the morning sky.

The southern chart is more complicated, because the Sun starts the calendar year not far west of the ecliptic's southernmost point, and Venus starts near the Sun and moves from side to side of it. So in each year Venus either passes that southernmost point twice, if starting on the Sun's west (2021, 2024, 2027), or not at all (2020, 2023, 2026), if starting on the Sun's east. The geometry allows Venus to pass that southernmost ecliptic point just once in 2022 and (barely) 2025.

Drawing these charts for the 5 synodic years, instead of the 8 calendar years, proves even more dense!

Pleiades passages

One of the benefits of a year when Venus is destined for its northernmost point is that on the way it visits the most famous of star clusters. The Pleiades (440 light years distant) are not as near to us as the Hyades (150), but being more compact they are the most readily noticed cluster in the sky. The"Seven Sisters" are actually five and their parents, the Titan Atlas and his wife Pleione. Besides the 6 or 7 or slightly more stars that various naked eyes see, there are over a thousand fainter cluster members. In deep photographs, they are gorgeously swathed in reflection nebulae.

As Venus's eight-year cycle evolves, the arc through the Pleiades shifts westward. Each time, the slope of the course lies about a tenth of a degree farther north, and will cut most centrally through the Pleiades in 2036.

Some of the stars–including many faint ones–get occulted by Venus–a visually exciting effect in the telescope. Calculating the exact band of places from which an occultation can be seen, and times of the star's disappearance into the dark side and reappearance from the sunlit side, is complex. If you want to try to see an occultation, prepare in advance with up-to-date information, including predictions and advice on observing and reporting, from IOTA, the International Occultation Timing Association, https://occultations.org/.

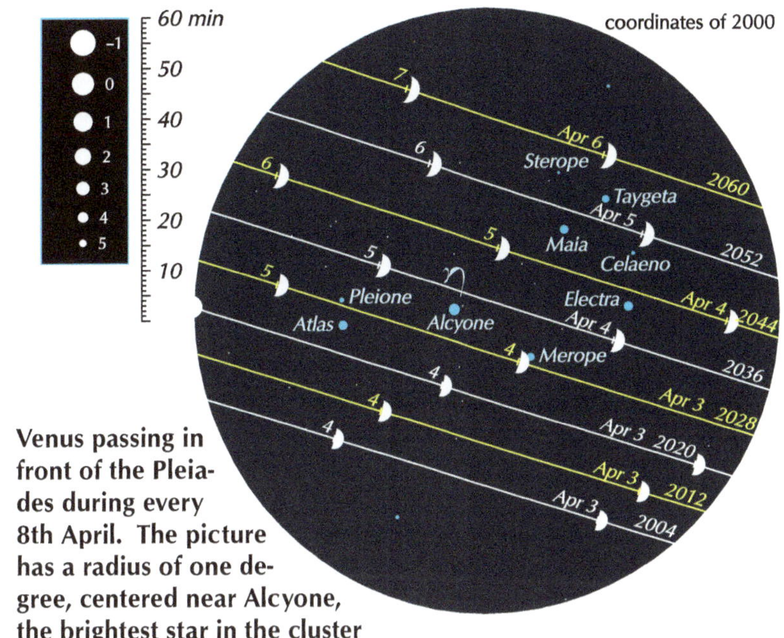

Venus passing in front of the Pleiades during every 8th April. The picture has a radius of one degree, centered near Alcyone, the brightest star in the cluster and the only one with a Greek-letter designation (η). The scale is indicated by the bar at the left, showing one degree divided into 60 minutes. The planet's sunlit face, shown at 0h Universal Time of each day, is 10 times larger than scale. Because the interval between years is 8 (they are all leap years), the successive positions for e.g. April 4 are evenly spaced. About two days before the date of each first appearance of Venus in this diagram, Venus joins the Pleiades in a low-power telescope field; a breath-taking sight.

Deduke men a selanna
kai Plêiades, mesai de
nuktes, para d'erkhet' ôra,
egô de mona kateudô.

The moon has set, and the Pleiades
in midnight have gone down,
and another hour goes crawling by
and I lie down alone.

–Sappho, who lived about 630-570 BC on Lesbos. Her Greek is the Aeolic dialect of that island: a selanna, "the moon," classical hê selênê.

The rainy Pleiads wester,
Orion plunges prone,
The stroke of midnight ceases
And I lie down alone.
The rainy Pleiads wester
And seek beyond the sea
The head that I shall dream of,
That shall not dream of me.

A.E. Housman, one of the 63 poems in A Shropshire Lad (published 1896). Housman was a formidable scholar of Greek. The Pleiades were probably the "sailing" stars (Greek *plein*, "to sail") because their heliacal rising–first visibility before dawn–signalled the spring start of the sailing season. More than two millennia later, that rising is more than two months later, and the Pleiades may be "rainy."

Venus and Pleiades. Photograph by James Weightman, Cirencester, England; taken on 2020 Apr. 4, a time when, as he said, astronomy was the "one activity not locked down!" Exposure 1 minute. The light of the planet, near easternmost elongation, overwhelms the stars.

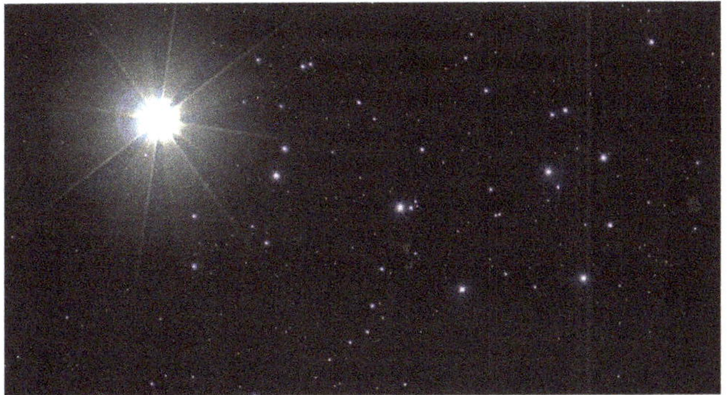

The cycle by apparition pairs

It's natural for us to want to see what Venus does in the course of the year we're in. But an alternative is to divide the cycle into Venus's performances. After all, January 1 interrupts a performance. The evening and morning performances, or apparitions, start at superior and inferior Sun-conjunction. It works better, descriptively, to divide the cycle in this way: ten apparitions, or five apparition pairs.

I am defining each apparition pair as an evening followed by a morning apparition. Thus it starts and ends with Venus at its most remote, behind the Sun (superior conjunction), and centers on the nearest moment (inferior conjunction). Though that central stage is also one of essential invisibility, it can be among the most interesting. It is then that Venus sometimes transits the Sun; or leaps high enough over it to be findable very near to the Sun-conjunction.

Each apparition pair has on average the length of Venus's synodic period, 584 days or about 1.6 years: the time in which we see it go around and return to the same apparent place in relation to the Sun.

I would have found a neater term than "apparition pair" if I could. "Hoop"? "Synody"? *Synod* ("togetherness of roads") has the historic meaning of a council of church people, usually to decide on some doctrinal issue, and *synody* is in the dictionary only as an obsolete variant of that.

"Drama", in two Acts? At inferior conjunction, Venus is an actress stepping through a door from an evening theater to a morning theater.

The following pages display five apparition pairs, constituting an eight-year cycle, plus a sixth pair to show how the cycle begins to repeat. On each of the following groups of four pages:

Charts

These small-scale equatorial charts span the 360° of the celestial sphere. The evening chart shows the path of Venus from superior conjunction through easternmost elongation to inferior conjunction, and the morning chart shows the travel back to the next superior conjunction.

Space views

These show the movements of Venus over six months centered on inferior conjunction. The viewpoint is at a distance of 6 astronomical units (a.u.) from the Sun, at a latitude of 40° north of the ecliptic plane and longitude 240° from the vernal equinox direction. This allows us to look down over the planets' orbits, and also to see the angle between the ecliptic and equatorial planes. On an imaginary sphere, 2 a.u. from the Sun, are shown the ecliptic (with dashes 2° long), the celestial equator (with ticks at intervals of one hour of right ascension), and the boundaries of the 12 zodiacal constellations. Earth moves in the ecliptic plane, and Venus in a plane tilted only 3.4° to it, so as seen from Earth Venus appears against the background of those constellations (except that part of the ecliptic, in the foreground, lies in Ophiuchus, north of Scorpius). Along the orbital paths, globes represent the planets at the start of each month. Their size is exaggerated 500 times, the Sun's only 5 times. When Venus is north of the ecliptic plane, its path is drawn with a thicker line. When it is in the morning sky (west of the Sun) as seen from the Earth, its course is shown in gray.

In 2020, Venus is at inferior conjunction on June 3 and descends to the south of the ecliptic plane on June 5.

If we were to draw the planets' paths for the whole year, that of Earth would be a closed circle, but Venus's path would overlap itself by about 63%. The 6-month paths enable us to compare the orbits more easily.

Horizon arcs

These show the positions of Venus above the western horizon at sunset, and the eastern horizon at sunrise, for latitude 40#d north (Philadelphia, and approximately Madrid, Rome, Istanbul, Beijing) and 35#d south (approximately Buenos Aires, Cape Town, Sydney, Auckland). The system is altazimuth (altitude from and azimuth along the horizon). The planet's disk, exaggerated 480 times in size, is drawn at days 1, 11, and 21 of each month; also (yellow) at the dates that are nearest to the moments of greatest elongation. (There is a westernmost elongation at 2020

Aug. 12, but it is near the end of the date, so the nearest morning view is on Aug. 13.) The disk's sunlit side faces toward the Sun, which is at different positions along the horizon. The dashed line is the celestial equator. Its angle to the horizon shows the drastic difference between views from northern and southern lands. For Earth's equator, the celestial equator would be vertical. To imagine the difference for latitude 50#d north, tilt the equator a little down; really, rotate the horizon a little upward around the west or east point. We give the sunset view on the left so as better to visualize the whole apparition pair, with inferior conjunction in the middle.

Tables of dates

Columns are: right ascension (in hours, minutes, seconds) and declination (in degrees and minutes), for epoch 2000; heliocentric and geocentric distance (distance from Sun and Earth) in astronomical units; elongation (negative meaning westward, in the morning sky); magnitude; angular diameter, in seconds; illuminated fraction (how much of the disk's width is sunlit); declination difference from the Sun (when negative, it is more helpful for south-hemisphere observers).

An hour may be rounded to "24" because the instant falls in the last hour of the UT day.

Events not spelled out are:
"sup.conj.": superior conjunction
"inf.conj.": inferior conjunction
"MAX.ELONG.": greatest elongation
"max.decl.": greatest declination
"max.lat.": greatest ecliptic latitude
"max.illum.area": greatest illuminated extent
"on equat.": crosses celestial equator
"stat.in long.": stationary in ecliptic longitude
"stat.in r.a.": stationary in right ascension
">retr": goes retrograde
">dir.": goes direct

Horizon scenes

The sky scene are at the evening closest to the instant of easternmost elongation and the morning closest to westernmost elongation. They are for a U.S. location (40° north, 90° west), but for other mid-northern regions there will be little difference in the positions of the moving bodies except the Moon.

Arrows through planets and Sun show their movement from 2 days before to 2 days after the picture date, in relation to the celestial background. The disks of Venus and Saturn are drawn at 150 times their true size. A broad arrow shows how far everything in the sky moves, parallel to the celestial equator, in an hour. The soft band sometimes in the picture is the Milky Way. This and many other features shown are for the mind's eye only, being too faint to be naked-eye visible in twilight or low to the horizon or at all. Some interesting points may appear in the pictures, such as "Earth's direction travel" and "antapex of Earth's way," to and from which Earth is traveling in its orbit. The projection is radial from a point 5° below the west or east point on the horizon, yielding a convex horizon and thus giving a sense that we are ourselves on a planet.

Graphs of favorability

From these graphs you can find the times when Venus will be highest above your sunset and sunrise horizons; and you can see how this varies for northern and southern parts of the Earth, and for the cycle.

Gray areas represent evening apparitions (eastward elongation); blue, morning apparitions (westward). The top figures are the maximum elongations (angular distances the planet attains from the Sun), reached at the top dates shown beneath. Curves show the altitude of the planet above the horizon at sunset or sunrise, for latitude 40° north (thick line) and 35° south (thin), with maxima reached at the parenthesized dates below (40° north bold).

For example, Venus reaches a greatest elongation of 46.1° on 2020 Mar. 24, and a greatest sunset altitude of 43.9° on Mar. 26 for 40° north, but for 35° south the greatest sunset altitude was only 25.0°, back on 2019 Dec. 25.

Inspection of the thick and thin curves will show which season of each year will give you the highest sunset or sunrise Venus, and the contrasts for latitude: which apparitions favor one or the other hemisphere.

The apparitions of a pair lean together: there is a narrow ditch between the evening and morning mountains. That's because Venus makes its rapid swoop between us and the Sun–from easternmost elongation through inferior conjunction to westernmost.

2019-2021 apparition pair, Taurus-centered

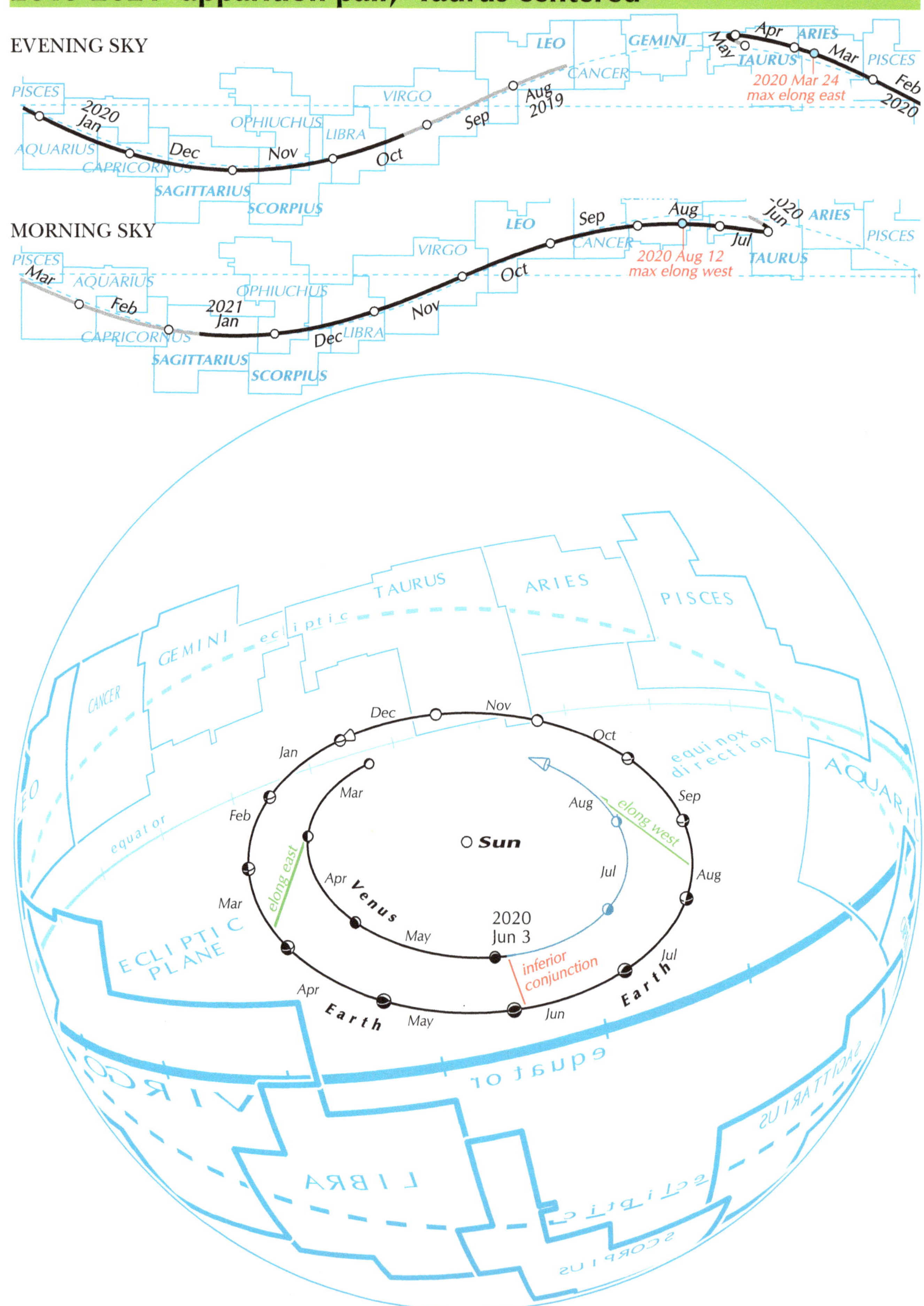

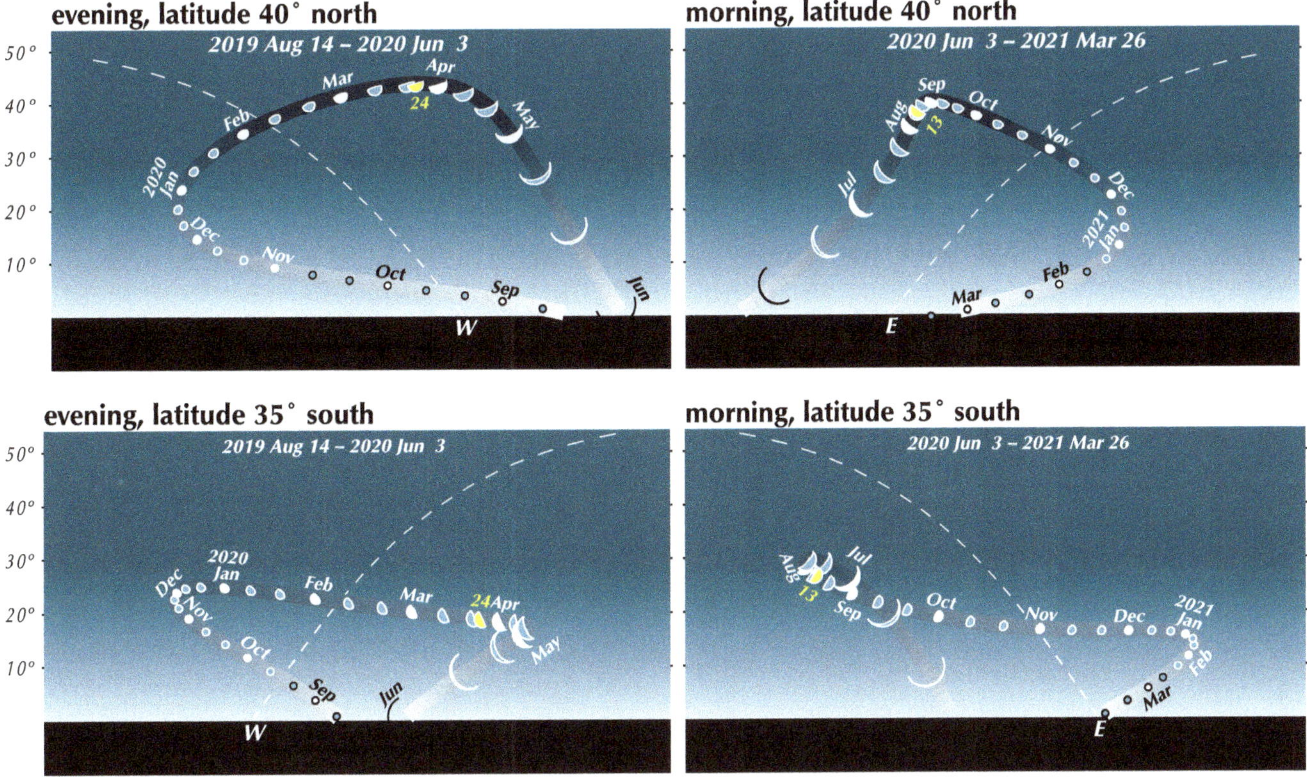

2019 Aug 14 **superior conjunction, in Leo**
2020 Mar 24 **easternmost elongation, in Aries**
2020 Jun 3 **inferior conjunction, in Taurus**
2020 Aug 12 **westernmost elongation, in Gemini**
2021 Mar 26 **superior conjunction, in Pisces**

This type of apparition pair in the cycle of five serves as an excellent first sample of Venus's display. It is almost symmetrical around its midpoint, the inferior conjunction near the middle of 2020. Hence the striking butterfly shape (for northern latitudes) of the sky scenes for the whole apparition pair and for the year 2020.

The cause of the symmetry is that the middle moment, the inferior conjunction, is just two days before Venus is at the descending node of its orbit. Nowhere else in the cycle is there such a close coincidence of Sun-conjunction and node. (The second closest is at the end of 2025: descending node 19 days before superior conjunction.)

This coincidence geometrically engineers the apparition pair's symmetry and its other very special features:

–At the inferior conjunction itself, the passage closest to the Sun in the whole cycle, making possible the transits–not this time, but in 2012 and 2004.

–Passage across the Pleiades, on Apr. 3 and 4.

–The northernmost declination reached by Venus in the whole cycle. 27.82°, on 2020 May 4.

The evening apparition begins with Venus's emergence from behind the Sun in Leo, so over the next months it moves out ahead and southward of the Sun, till November, then northward. The evening maximum of elongation is high northward.

Then the retrograde motion takes place in Taurus, an S-shape down across the northernmost part of the ecliptic.

But the morning westernmost elongation is still quite northerly. Then comes the long circling away from us, southward ahead of the Sun till the beginning of 2021, then northward with Venus at its most remote.

2019		RA (2000)	decl	hedis	gedis	elo	mag	dia″	lum	dsu
Aug 14 6	sup.conj.with Sun	9 34 52	15 44	0.719	1.731	1	-3.9	9.7	100	1
Aug 15 12	brightest	9 40 60	15 15	0.719	1.731	1	-3.9	9.7	100	1
Aug 24 13	0.3°N of Mars	10 24 16	11 28	0.719	1.726	3	-3.9	9.7	100	0
Aug 30 1	max.lat.north	10 50 5	8 57	0.719	1.722	5	-3.9	9.7	100	0
Sep 13 22	0.3°N of Mercury	11 58 6	1 38	0.721	1.701	8	-3.9	9.8	99	-2
Sep 16 22	on equat.,to sou.	12 11 45	0 6	0.721	1.695	9	-3.9	9.9	99	-2
Oct 25 2	descending node	15 10 40	-17 45	0.726	1.590	19	-3.9	10.5	95	-6
Oct 30 8	2.7°N of Mercury	15 37 7	-19 36	0.727	1.571	20	-3.9	10.7	94	-6
Nov 24 14	1.4°s of Jupiter	17 51 3	-24 40	0.728	1.468	26	-3.9	11.4	90	-4
Nov 28 17	max.declin.south	18 13 31	-24 47	0.728	1.449	27	-3.9	11.6	89	-3
Nov 28 18	aphelion	18 13 55	-24 47	0.728	1.449	27	-3.9	11.6	89	-3
Dec 11 5	1.8°s of Saturn	19 21 19	-23 50	0.728	1.389	30	-4.0	12.1	87	-1
Dec 13 12	1.1°s of Pluto	19 33 35	-23 27	0.728	1.377	31	-4.0	12.2	86	0
Dec 20 23	max.lat.south	20 12 47	-21 49	0.727	1.338	32	-4.0	12.5	85	2

2020		RA (2000)	decl	hedis	gedis	elo	mag	dia″	lum	dsu
Jan 1 0		21 8 40	-18 20	0.726	1.278	34	-4.0	13.1	82	5
Jan 27 19	0.1°s of Neptune	23 12 34	-6 15	0.723	1.117	40	-4.1	15.0	75	12
Feb 8 16	on equat.,to nor.	0 3 19	0 -6	0.721	1.039	42	-4.1	16.1	71	15
Feb 15 6	ascending node	0 30 44	3 19	0.721	0.994	43	-4.2	16.8	69	16
Mar 9 15	2.4°N of Uranus	2 6 48	14 46	0.719	0.825	45	-4.3	20.3	59	19
Mar 20 3	perihelion	2 49 21	19 5	0.718	0.746	46	-4.3	22.5	54	19
Mar 24 22	MAX.ELONG.EAS	3 8 33	20 48	0.718	0.709	46	-4.4	23.6	51	19
Mar 27 1	dichotomy	3 16 54	21 31	0.719	0.692	46	-4.4	24.2	50	19
Apr 10 18	max.lat.north	4 12 15	25 25	0.719	0.579	45	-4.5	28.9	41	17
Apr 28 1	max.illum.area	5 3 34	27 38	0.721	0.450	40	-4.5	37.2	28	13
Apr 28 15	brightest	5 4 47	27 40	0.721	0.446	40	-4.5	37.5	27	13
May 4 0	max.declin.north	5 14 45	27 48	0.722	0.409	37	-4.5	40.9	22	12
May 13 7	stat.in long>retr	5 22 3	27 24	0.723	0.353	29	-4.4	47.4	13	9
May 13 10	stat.in r.a.>retr	5 22 3	27 23	0.723	0.352	29	-4.4	47.5	13	9
May 22 8	0.9°N of Mercury	5 15 9	26 8	0.724	0.312	19	-4.2	53.7	5	6

============

2020		RA (2000)	decl	hedis	gedis	elo	mag	dia″	lum	dsu
Jun 3 18	inf.conj.with Sun	4 47 24	22 53	0.726	0.289	0	-0.9	58.0	0	0
Jun 5 19	descending node	4 42 7	22 14	0.726	0.289	-3	-3.8	57.9	0	0
Jun 24 18	stat.in r.a.>dir.	4 15 10	17 40	0.728	0.351	-29	-4.4	47.7	13	-6
Jun 25 7	stat.in long>dir.	4 15 11	17 37	0.728	0.354	-29	-4.4	47.3	13	-6
Jul 8 12	brightest	4 28 20	17 16	0.728	0.437	-39	-4.5	38.3	25	-5
Jul 10 8	max.illum.area	4 31 49	17 21	0.728	0.450	-40	-4.5	37.2	27	-5
Jul 10 14	aphelion	4 32 23	17 22	0.728	0.452	-40	-4.5	37.1	27	-5
Aug 1 15	max.lat.south	5 36 25	19 20	0.727	0.618	-45	-4.4	27.1	43	2
Aug 12 21	dichotomy	6 19 36	20 2	0.726	0.706	-46	-4.3	23.7	50	5
Aug 12 24	MAX.ELONG.WES	6 20 8	20 2	0.726	0.707	-46	-4.3	23.7	50	5
Sep 26 22	ascending node	9 40 49	13 54	0.721	1.045	-41	-4.1	16.0	70	16
Oct 30 23	perihelion	12 14 28	0 9	0.718	1.265	-34	-4.0	13.2	81	14
Oct 31 2	on equat.,to sou.	12 14 55	0 7	0.718	1.266	-34	-4.0	13.2	81	14
Nov 21 10	max.lat.north	13 52 36	-9 39	0.719	1.384	-30	-4.0	12.1	87	10

2021		RA (2000)	decl	hedis	gedis	elo	mag	dia″	lum	dsu
Jan 1 0		17 17 16	-22 24	0.724	1.560	-20	-3.9	10.7	94	1
Jan 12 6	max.declin.south	18 18 26	-23 10	0.726	1.598	-18	-3.9	10.5	96	-2
Jan 16 12	descending node	18 41 32	-23 5	0.726	1.611	-17	-3.9	10.4	96	-2
Jan 28 19	0.7°N of Pluto	19 48 1	-21 37	0.727	1.644	-14	-3.9	10.2	97	-4
Feb 6 5	0.4°s of Saturn	20 32 20	-19 39	0.728	1.664	-12	-3.9	10.1	98	-4
Feb 11 12	0.4°s of Jupiter	20 59 33	-18 4	0.728	1.675	-11	-3.9	10.0	98	-4
Feb 12 17	4.8°s of Mercury	21 5 37	-17 41	0.728	1.678	-10	-3.9	10.0	99	-4
Feb 20 8	aphelion	21 43 49	-14 54	0.728	1.691	-8	-3.9	9.9	99	-4
Mar 14 1	0.4°s of Neptune	23 26 39	-5 9	0.727	1.717	-3	-3.9	9.7	100	-3
Mar 14 8	max.lat.south	23 28 2	-5 0	0.727	1.718	-3	-3.9	9.7	100	-2
Mar 24 3	on equat.,to nor.	0 12 48	0 -6	0.726	1.723	1	-3.9	9.7	100	-1

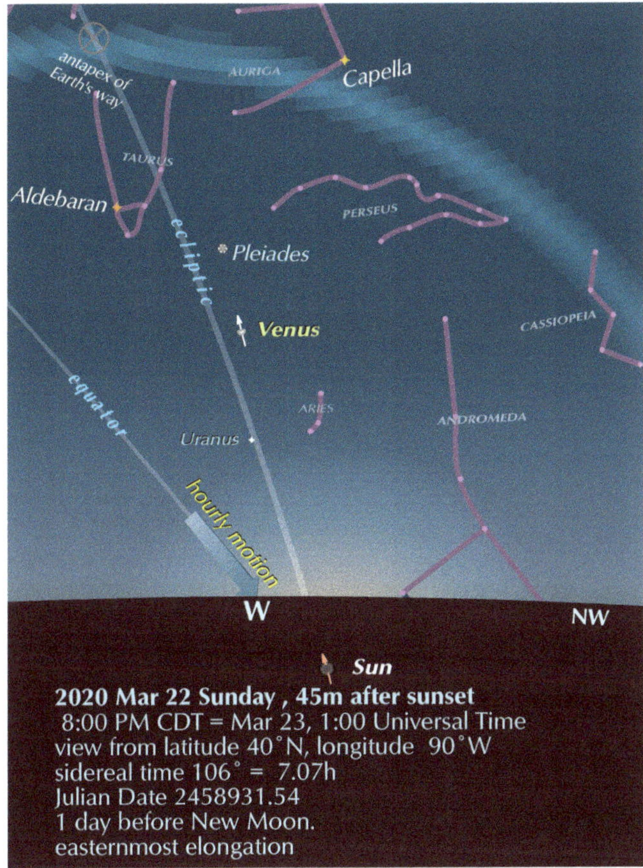

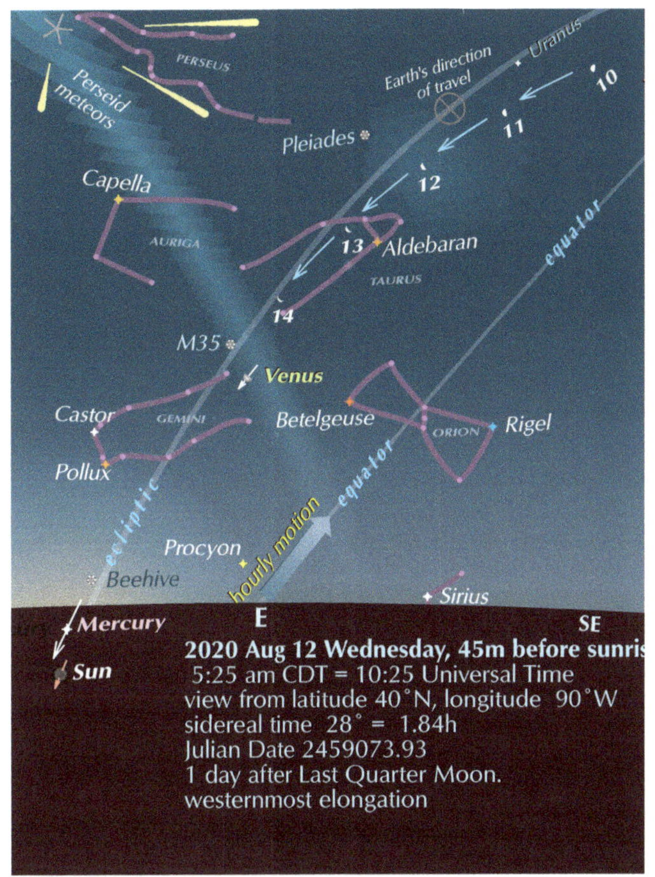

The morning elongation maximum, being on August 12, comes near the probable peak of the Perseids, the most reliable of the major annual meteor showers. Any "shooting star" whose path can be traced back to the radiant point near the north end of Perseus is Perseid.

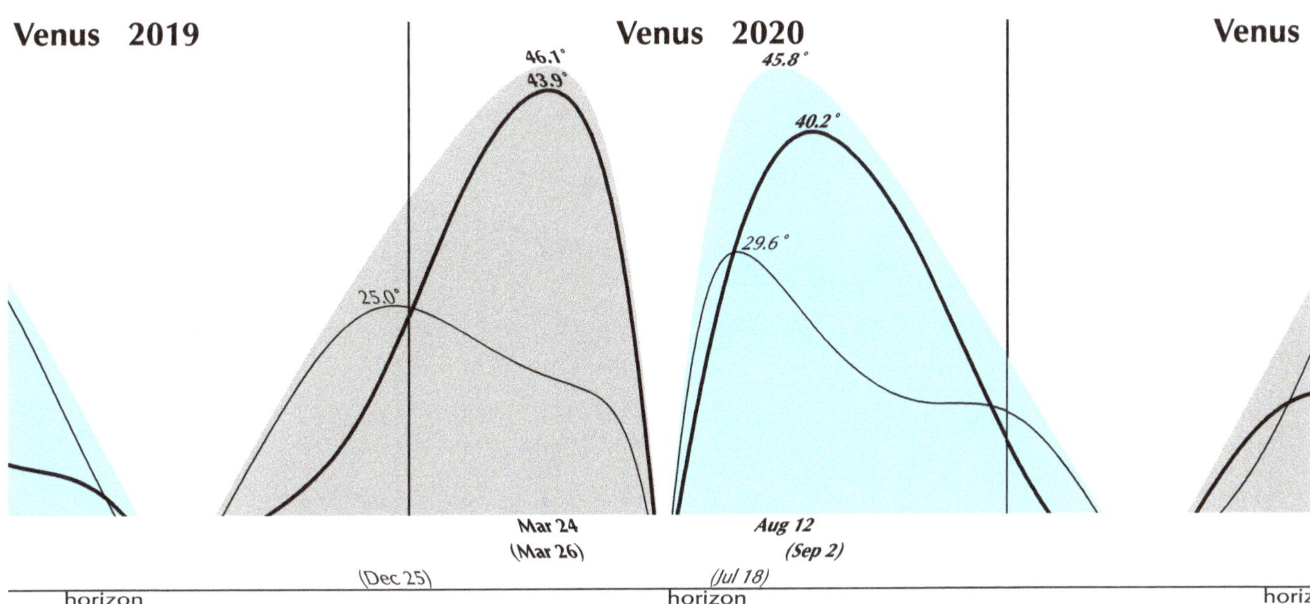

2021-2022 apparition pair, Sagittarius-centered

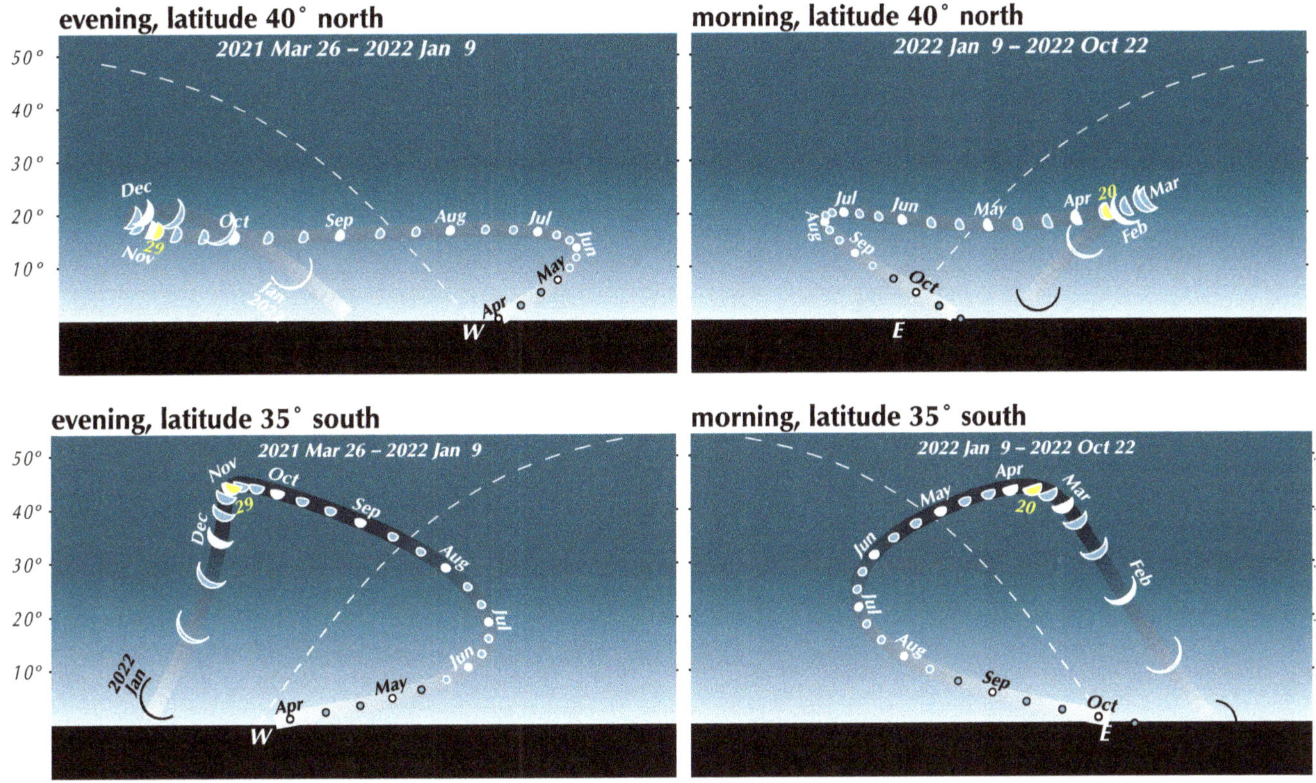

2021 Mar 26	**superior conjunction, in Pisces**
2021 Oct 29	**easternmost elongation, in Ophiuchus**
2022 Jan 9	**inferior conjunction, in Sagittarius**
2022 Mar 20	**westernmost elongation, in Capricornus**
2022 Oct 22	**superior conjunction, in Virgo**

This type of apparition pair is almost the opposite of the preceding one. Venus at its best showing in the evening is about as far south as it can be, because near that time it reaches its southernmost point in the whole cycle.

It's fated to be that way because of where superior conjunction happens: close to the zero of the sky, the equinox point in Pisces. This means that half of its synodic orbit later Venus will have gone around the sky about one and a quarter times to where we are looking southward toward it.

Its course out into the evening sky is northward until May-June, then for five months all along the southward slope of the ecliptic, to the evening elongation. This is well south of the southernmost ecliptic, in the part of Ophiuchus that "should" belong to Scorpius. And shortly afterward, on 2021 Nov. 6, Venus is at its southernmost declination in the whole eight-year cyle.

The retrograde path, in December and January, is a northward Z across the ecliptic in Sagittarius. At inferior conjunction, Venus is 5° north of the Sun, quite high but not as high as in 2025.

So Venus comes to a less deep-southerly morning extreme of elongation (on the northern edge of Capricornus). The journey away from us is a long arch, northward for three months, then southward for the remaining four.

2021		RA (2000)	decl	hedis	gedis	elo	mag	dia"	lum	dsu
Mar 26 6	sup.conj.with Sun	0 22 26	0 57	0.726	1.723	1	-3.9	9.7	100	-1
Mar 29 2	brightest	0 35 20	2 23	0.726	1.723	2	-3.9	9.7	100	-1
Apr 22 23	0.3°s of Uranus	2 30 41	14 9	0.723	1.709	7	-3.9	9.8	99	2
Apr 26 9	1.3°s of Mercury	2 47 3	15 33	0.722	1.705	8	-3.9	9.8	99	2
May 9 15	ascending node	3 52 57	20 14	0.721	1.681	11	-3.9	10.0	98	3
May 29 5	0.4°N of Mercury	5 35 34	24 6	0.719	1.629	17	-3.9	10.3	96	2
Jun 5 12	max.declin.north	6 14 34	24 26	0.719	1.603	19	-3.9	10.4	95	2
Jun 12 18	perihelion	6 53 19	24 9	0.718	1.575	21	-3.9	10.6	93	1
Jul 4 3	max.lat.north	8 44 4	19 49	0.719	1.476	26	-3.9	11.3	89	-3
Jul 13 7	0.5°N of Mars	9 28 41	16 38	0.720	1.427	28	-3.9	11.7	87	-5
Aug 17 11	on equat.,to sou.	12 5 29	0 7	0.725	1.209	37	-4.0	13.8	77	-13
Aug 29 5	descending node	12 55 16	-5 54	0.726	1.128	39	-4.0	14.8	74	-15
Oct 3 1	aphelion	15 27 24	-21 12	0.728	0.872	45	-4.2	19.2	61	-17
Oct 25 1	max.lat.south	17 8 15	-26 24	0.727	0.704	47	-4.4	23.8	52	-14
Oct 28 14	dichotomy	17 24 19	-26 48	0.727	0.677	47	-4.4	24.7	50	-14
Oct 29 21	MAX.ELONG.EAS	17 29 56	-26 55	0.727	0.667	47	-4.4	25.1	49	-13
Nov 6 16	max.declin.south	18 3 50	-27 14	0.726	0.608	47	-4.5	27.5	45	-11
Dec 4 7	max.illum.area	19 37 1	-24 12	0.723	0.408	40	-4.7	41.0	26	-2
Dec 7 16	brightest	19 43 28	-23 32	0.722	0.386	38	-4.7	43.3	23	-1
Dec 11 19	0.1°N of Pluto	19 49 17	-22 41	0.722	0.361	35	-4.7	46.4	19	0
Dec 18 11	stat.in r.a.>retr	19 53 5	-21 17	0.721	0.324	30	-4.6	51.7	13	2
Dec 19 11	stat.in long>retr	19 52 60	-21 5	0.721	0.319	29	-4.6	52.4	12	2
Dec 20 8	ascending node	19 52 46	-20 54	0.721	0.315	28	-4.6	53.2	11	3
Dec 23 13	2.5°N of Pluto	19 50 48	-20 14	0.720	0.301	24	-4.5	55.7	9	3
Dec 29 1	4.2°N of Mercury	19 43 17	-19 9	0.720	0.281	17	-4.4	59.5	4	4

2022		RA (2000)	decl	hedis	gedis	elo	mag	dia"	lum	dsu
Jan 1 0		19 37 16	-18 37	0.719	0.274	13	-4.3	61.1	2	4

============

2022		RA (2000)	decl	hedis	gedis	elo	mag	dia"	lum	dsu
Jan 9 1	inf.conj.with Sun	19 17 3	-17 21	0.719	0.266	-5	-4.0	63.0	0	5
Jan 23 6	perihelion	18 47 48	-16 12	0.718	0.297	-22	-4.5	56.3	7	3
Jan 29 8	stat.in r.a.>dir.	18 44 36	-16 12	0.719	0.325	-29	-4.6	51.5	13	2
Jan 29 9	stat.in long>dir.	18 44 36	-16 12	0.719	0.325	-29	-4.6	51.5	13	2
Feb 9 14	brightest	18 54 29	-16 35	0.719	0.391	-38	-4.6	42.8	23	-2
Feb 12 22	max.illum.area	19 0 46	-16 44	0.719	0.413	-40	-4.6	40.5	26	-3
Feb 13 1	6.6°N of Mars	19 0 59	-16 44	0.719	0.414	-40	-4.6	40.5	26	-3
Feb 13 20	max.lat.north	19 2 41	-16 46	0.719	0.419	-40	-4.6	39.9	27	-4
Mar 5 4	5.7°N of Pluto	20 0 15	-16 44	0.721	0.560	-46	-4.5	29.9	41	-11
Mar 12 15	4.0°N of Mars	20 28 10	-16 5	0.722	0.617	-46	-4.5	27.1	45	-13
Mar 20 9	MAX.ELONG.WES	20 59 0	-14 58	0.723	0.678	-47	-4.4	24.7	50	-15
Mar 21 6	dichotomy	21 2 29	-14 49	0.723	0.685	-47	-4.4	24.5	50	-15
Mar 29 13	2.2°N of Saturn	21 36 50	-13 3	0.725	0.750	-46	-4.3	22.3	54	-16
Apr 10 21	descending node	22 28 45	-9 31	0.726	0.846	-45	-4.2	19.8	60	-18
Apr 27 19	0.0°s of Neptune	23 40 23	-3 21	0.728	0.975	-43	-4.1	17.2	66	-17
Apr 30 19	0.2°s of Jupiter	23 53 4	-2 9	0.728	0.997	-43	-4.1	16.8	68	-17
May 5 18	on equat.,to nor.	0 14 7	0 -6	0.728	1.034	-42	-4.1	16.2	69	-16
May 15 11	aphelion	0 55 43	3 59	0.728	1.105	-40	-4.0	15.2	73	-15
Jun 6 18	max.lat.south	2 34 33	13 5	0.727	1.257	-35	-4.0	13.3	80	-10
Jun 11 13	1.6°s of Uranus	2 56 48	14 50	0.727	1.288	-34	-3.9	13.0	81	-8
Jul 22 3	max.declin.north	6 20 6	22 52	0.722	1.512	-24	-3.9	11.1	91	3
Aug 2 1	ascending node	7 17 17	22 15	0.720	1.560	-22	-3.9	10.7	93	4
Sep 4 20	perihelion	10 7 24	12 50	0.718	1.667	-13	-3.9	10.0	98	6
Sep 26 1	3.8°N of Mercury	11 46 11	3 3	0.719	1.703	-7	-3.9	9.8	99	4
Sep 26 13	max.lat.north	11 48 25	2 49	0.719	1.704	-7	-3.9	9.8	99	4
Oct 1 22	on equat.,to sou.	12 13 5	0 7	0.720	1.709	-6	-3.9	9.8	100	3
Oct 19 16	brightest	13 34 49	-8 39	0.722	1.717	-1	-3.9	9.7	100	1

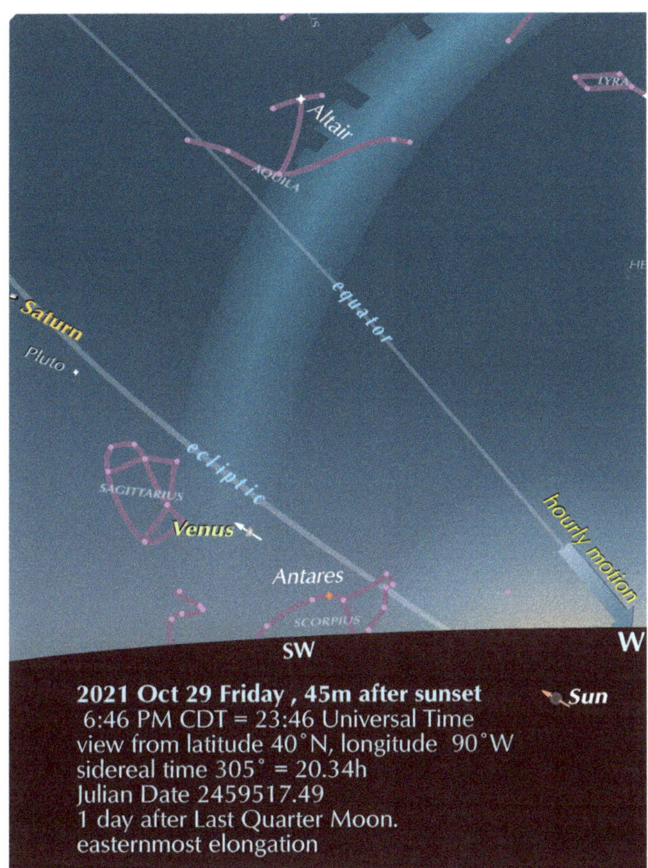
Altair
AQUILA
LYRA
equator
Saturn
Pluto
ecliptic
SAGITTARIUS
Venus
hourly motion
Antares
SCORPIUS
SW
W
Sun
2021 Oct 29 Friday , 45m after sunset
6:46 PM CDT = 23:46 Universal Time
view from latitude 40˚N, longitude 90˚W
sidereal time 305˚ = 20.34h
Julian Date 2459517.49
1 day after Last Quarter Moon.
easternmost elongation

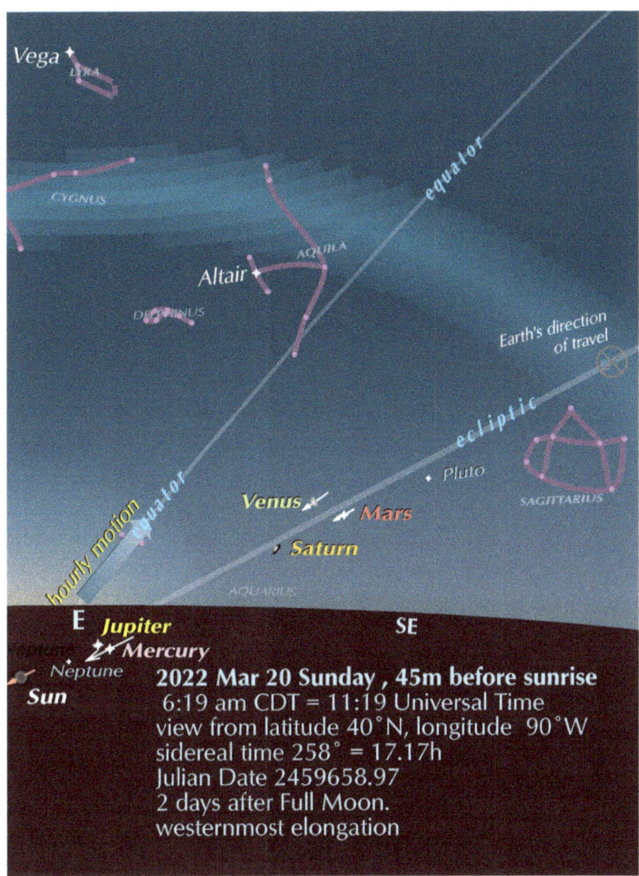
Vega
LYRA
CYGNUS
equator
AQUILA
Altair
DELPHINUS
Earth's direction of travel
ecliptic
Pluto
SAGITTARIUS
Venus
Mars
Saturn
hourly motion
equator
AQUARIUS
E
Jupiter
Mercury
SE
Neptune
Sun
2022 Mar 20 Sunday , 45m before sunrise
6:19 am CDT = 11:19 Universal Time
view from latitude 40˚N, longitude 90˚W
sidereal time 258˚ = 17.17h
Julian Date 2459658.97
2 days after Full Moon.
westernmost elongation

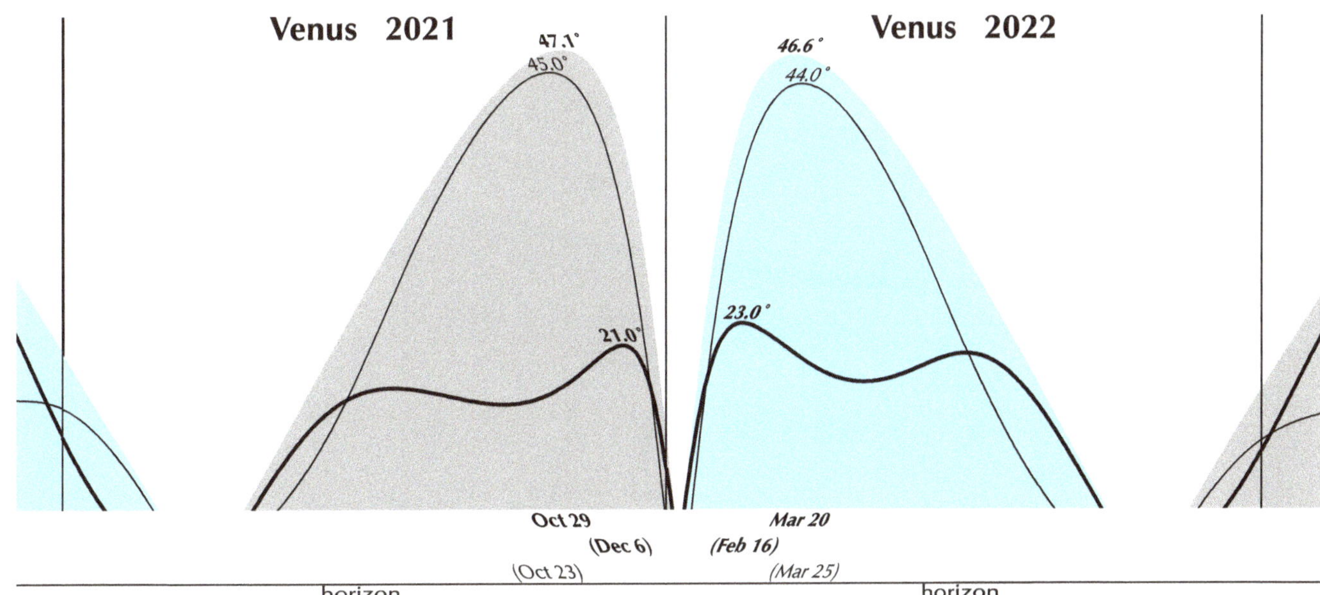
Venus 2021
Venus 2022
45.0°
21.0°
46.6°
44.0°
23.0°
Oct 29
(Dec 6)
(Oct 23)
Mar 20
(Feb 16)
(Mar 25)
horizon
horizon

2022-2024 apparition pair, Leo-centered

EVENING SKY

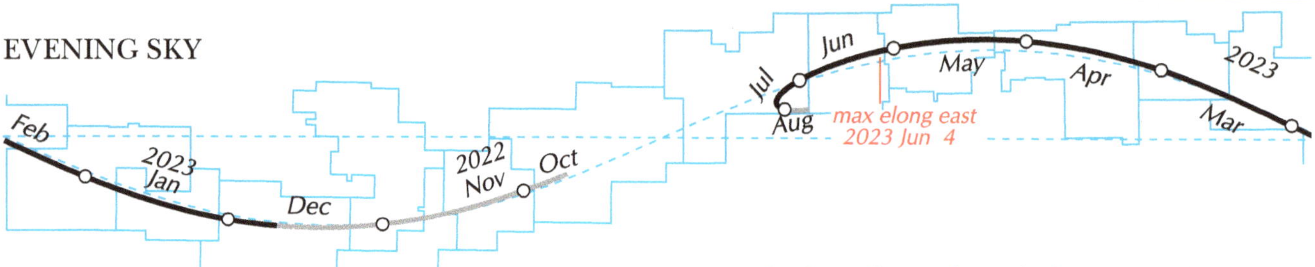

MORNING SKY

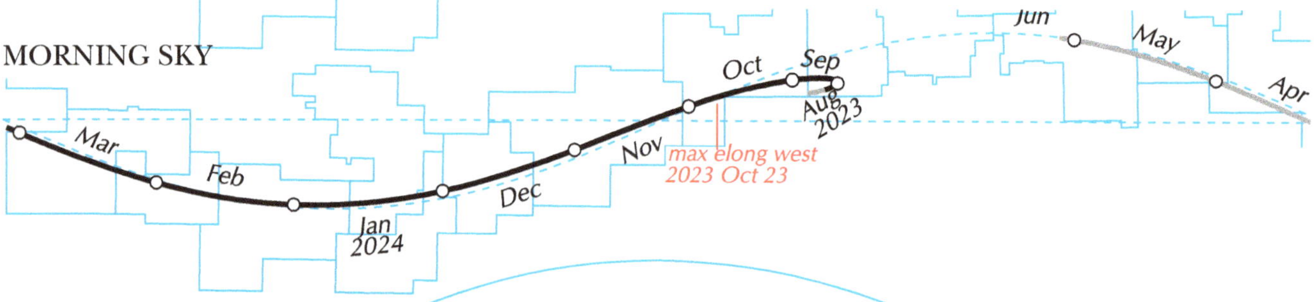

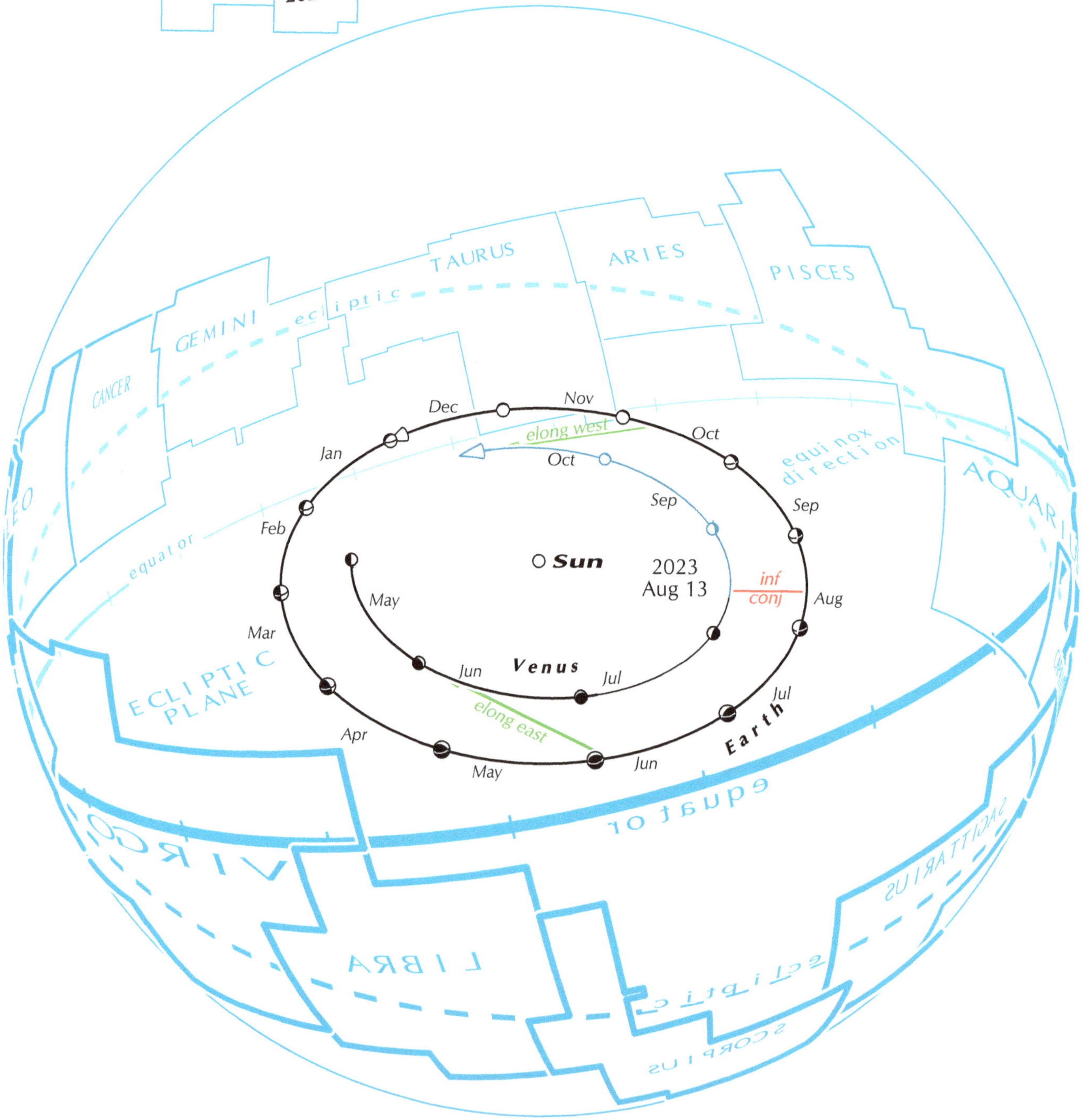

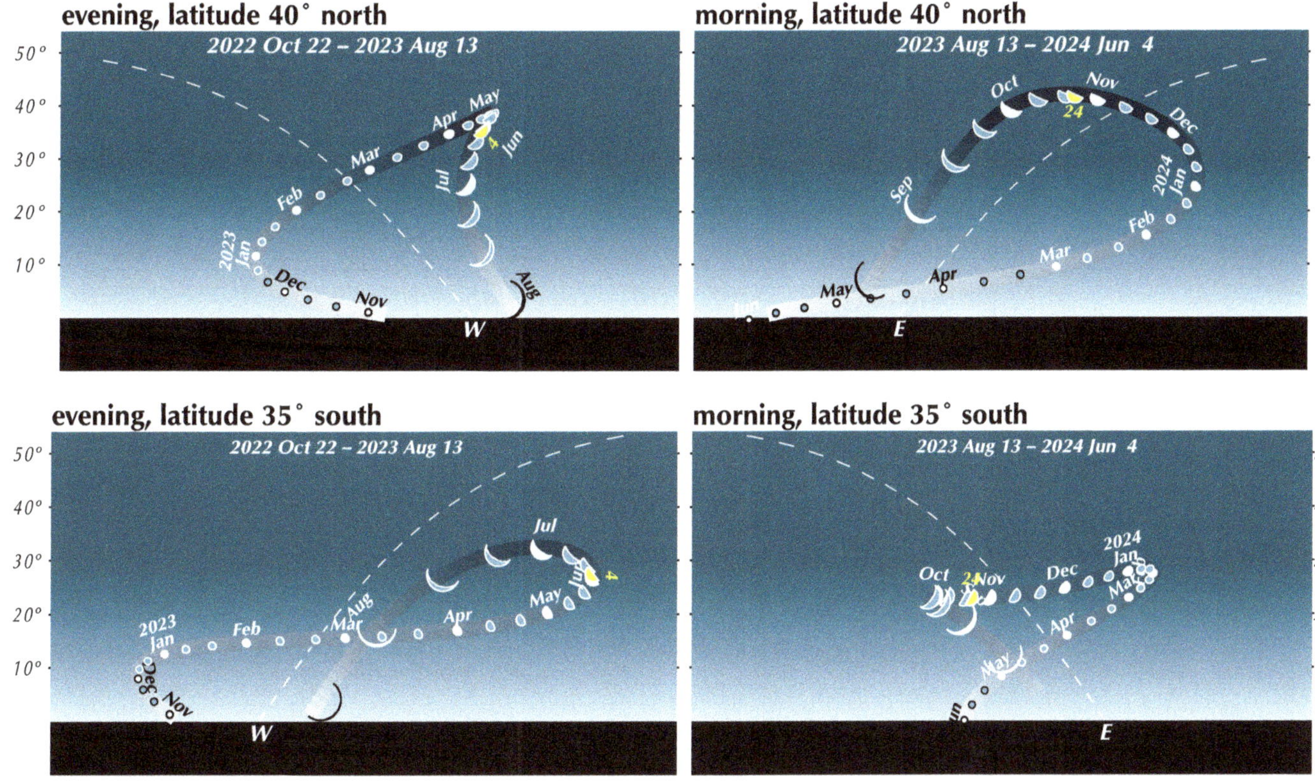

2022 Oct 22	**superior conjunction, in Virgo**
2023 Jun 4	**easternmost elongation, in Cancer**
2023 Aug 13	**inferior conjunction, in Leo**
2023 Oct 23	**westernmost elongation, in Leo**
2024 Jun 4	**superior conjunction, in Taurus**

The evening apparition is modestly good for the northern hemisphere, moderately poor for the southern.

Starting from behind the Sun, near Spica in Virgo, Venus moves out at first slightly southward, then from December northward, and is northernmost in May.

It has traveled southward to Cancer by the time of easternmost elongation. A few days after that moment, it passes in front of the Beehive Cluster in Cancer.

The retrograde motion is a deep southward loop in Leo and Cancer, almost carrying Venus down to the group of stars that forms the "head" of Hydra. So at inferior conjunction Venus passes as much as 7° south of the Sun.

On the morning side, the westernmost elongation (almost into the corner of the bare little constellation Sextans) is still just in the northern celestial hemisphere. Receding, Venus moves southward till January, then northward till it disappears behind the Sun in Taurus.

2022		RA (2000)	decl	hedis	gedis	elo	mag	dia″	lum	dsu
Oct 22 21	sup.conj.with Sun	13 49 48	-10 10	0.722	1.717	1	-3.9	9.7	100	1
Nov 21 14	descending node	16 18 9	-21 21	0.726	1.694	7	-3.9	9.9	99	-1
Nov 22 17	1.4°N of Mercury	16 23 58	-21 38	0.726	1.693	8	-3.9	9.9	99	-1
Dec 13 2	max.declin.south	18 14 52	-24 13	0.728	1.656	13	-3.9	10.1	98	-1
Dec 26 3	aphelion	19 26 11	-23 12	0.728	1.624	16	-3.9	10.3	97	03
Dec 29 9	1.4°s of Mercury	19 43 42	-22 38	0.728	1.615	17	-3.9	10.4	96	1

2023		RA (2000)	decl	hedis	gedis	elo	mag	dia″	lum	dsu
Jan 1 0		19 57 39	-22 5	0.728	1.607	17	-3.9	10.4	96	1
Jan 1 9	0.9°N of Pluto	19 59 38	-22 0	0.728	1.606	17	-3.9	10.4	96	1
Jan 17 11	max.lat.south	21 22 21	-17 2	0.727	1.555	21	-3.9	10.8	94	4
Jan 22 20	0.4°s of Saturn	21 48 46	-14 53	0.727	1.536	22	-3.9	10.9	93	5
Feb 15 12	0.0°s of Neptune	23 38 56	-3 33	0.724	1.437	28	-4.0	11.6	89	9
Feb 22 3	on equat.,to nor.	0 8 26	0 -7	0.723	1.406	29	-4.0	11.9	87	10
Mar 2 11	0.5°N of Jupiter	0 45 26	4 13	0.722	1.364	31	-4.0	12.3	86	12
Mar 14 17	ascending node	1 40 18	10 25	0.720	1.297	33	-4.0	12.9	82	13
Mar 31 6	1.3°N of Uranus	2 56 18	17 44	0.719	1.196	37	-4.0	14.0	78	14
Apr 17 13	perihelion	4 19 34	23 18	0.718	1.080	40	-4.1	15.5	72	13
May 9 6	max.lat.north	6 6 33	26 5	0.719	0.919	44	-4.2	18.2	63	9
May 9 22	max.declin.north	6 9 57	26 5	0.719	0.913	44	-4.2	18.3	63	9
Jun 4 3	dichotomy	8 3 49	23 1	0.722	0.712	45	-4.3	23.5	50	1
Jun 4 11	MAX.ELONG.EAS	8 5 9	22 56	0.722	0.710	45	-4.3	23.6	50	0
Jul 4 7	descending node	9 39 10	14 2	0.726	0.474	41	-4.5	35.3	30	-9
Jul 7 20	max.illum.area	9 45 40	12 55	0.727	0.449	39	-4.5	37.3	27	-10
Jul 9 18	brightest	9 48 41	12 19	0.727	0.435	38	-4.5	38.5	25	-10
Jul 20 24	stat.in r.a.>retr	9 57 13	9 15	0.728	0.363	31	-4.4	46.1	15	-11
Jul 23 2	stat.in long>retr	9 56 53	8 48	0.728	0.352	29	-4.4	47.6	13	-11
Jul 26 13	5.3°s of Mercury	9 54 55	8 9	0.728	0.334	25	-4.3	50.1	10	-11
Aug 8 0	aphelion	9 33 54	7 10	0.728	0.293	11	-4.0	57.1	2	-9

2023		RA (2000)	decl	hedis	gedis	elo	mag	dia″	lum	dsu
Aug 13 11	inf.conj.with Sun	9 20 27	7 28	0.728	0.289	-8	-3.9	58.0	1	-7
Aug 30 3	max.lat.south	8 50 42	9 48	0.727	0.326	-25	-4.4	51.3	9	1
Sep 3 4	stat.in r.a.>dir.	8 49 24	10 22	0.727	0.345	-29	-4.4	48.6	13	36
Sep 4 1	stat.in long>dir.	8 49 28	10 29	0.727	0.349	-30	-4.5	47.9	14	3
Sep 18 12	brightest	9 6 15	11 31	0.725	0.436	-40	-4.5	38.4	27	10
Sep 19 7	max.illum.area	9 7 51	11 32	0.725	0.441	-40	-4.5	38.0	27	10
Oct 22 22	dichotomy	10 55 3	6 38	0.721	0.686	-46	-4.4	24.4	50	18
Oct 23 23	MAX.ELONG.WES	10 59 7	6 20	0.721	0.694	-46	-4.4	24.1	51	18
Oct 25 10	ascending node	11 4 49	5 54	0.721	0.705	-46	-4.4	23.7	51	18
Nov 11 5	on equat.,to sou.	12 12 19	0 8	0.719	0.831	-45	-4.3	20.1	60	17
Nov 28 13	perihelion	13 25 53	-6 45	0.718	0.957	-43	-4.2	17.5	67	14
Dec 19 22	max.lat.north	15 3 33	-14 58	0.719	1.104	-40	-4.1	15.2	74	8

2024		RA (2000)	decl	hedis	gedis	elo	mag	dia″	lum	dsu
Jan 1 0		16 2 27	-18 41	0.720	1.182	-37	-4.1	14.2	78	4
Jan 27 15	max.declin.south	18 21 21	-22 29	0.724	1.339	-32	-4.0	12.5	85	-4
Feb 13 24	descending node	19 53 28	-20 52	0.726	1.429	-28	-3.9	11.7	88	-8
Feb 17 21	2.7°N of Pluto	20 13 42	-20 4	0.727	1.448	-27	-3.9	11.6	89	-8
Feb 22 16	0.6°N of Mars	20 38 18	-18 52	0.727	1.471	-26	-3.9	11.4	90	-9
Mar 19 22	aphelion	22 46 31	-9 9	0.728	1.581	-20	-3.9	10.6	94	-9
Mar 22 2	0.3°N of Saturn	22 56 37	-8 10	0.728	1.589	-19	-3.9	10.5	95	-9
Apr 3 11	0.3°s of Neptune	23 53 23	-2 19	0.728	1.631	-16	-3.9	10.3	96	-8
Apr 7 23	on equat.,to nor.	0 13 50	0 -7	0.728	1.644	-15	-3.9	10.2	97	-7
Apr 10 20	max.lat.south	0 26 53	1 17	0.727	1.653	-14	-3.9	10.1	97	-7
Apr 18 23	2.0°s of Mercury	1 3 43	5 14	0.727	1.674	-12	-3.9	10.0	98	-6
May 18 9	0.5°s of Uranus	3 22 58	17 49	0.723	1.726	-5	-3.9	9.7	100	-2
May 23 9	0.2°N of Jupiter	3 48 4	19 26	0.722	1.731	-3	-3.9	9.7	100	-1

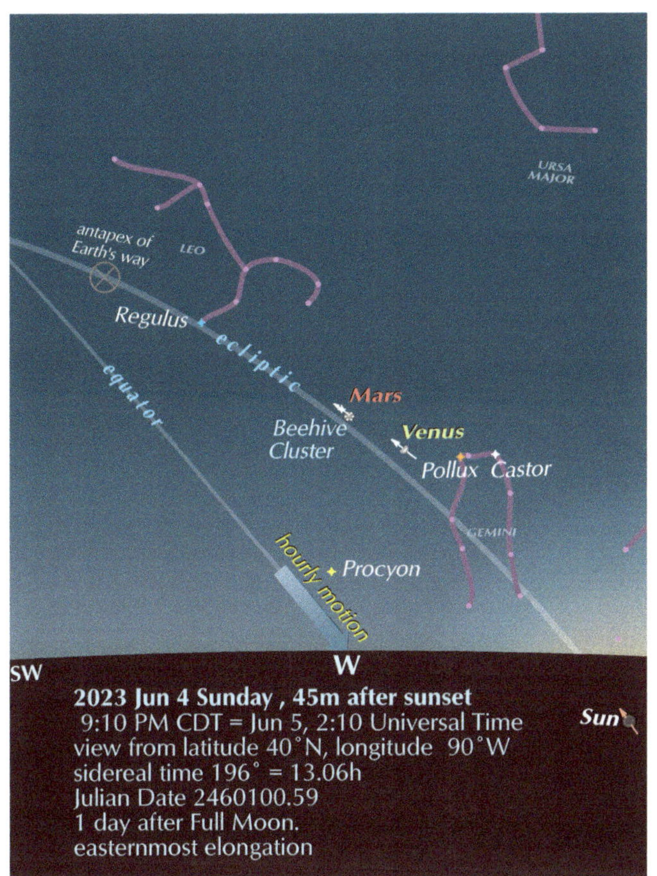
URSA MAJOR
antapex of Earth's way
LEO
Regulus
ecliptic
equator
Mars
Beehive Cluster
Venus
Pollux Castor
GEMINI
hourly motion
Procyon
SW
W
Sun
2023 Jun 4 Sunday , 45m after sunset
9:10 PM CDT = Jun 5, 2:10 Universal Time
view from latitude 40°N, longitude 90°W
sidereal time 196° = 13.06h
Julian Date 2460100.59
1 day after Full Moon.
easternmost elongation

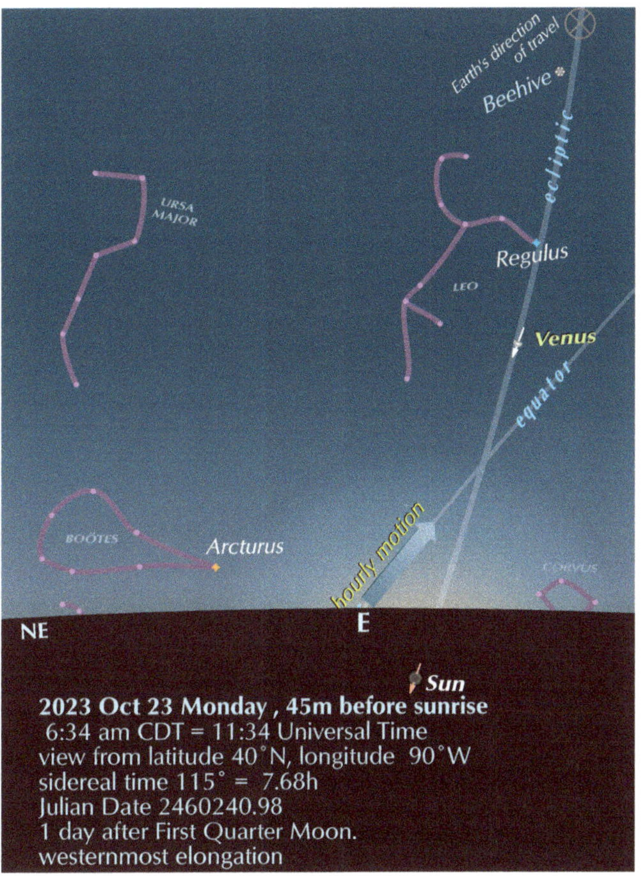
Earth's direction of travel
Beehive
ecliptic
URSA MAJOR
Regulus
LEO
Venus
equator
BOÖTES
Arcturus
hourly motion
CORVUS
NE
E
Sun
2023 Oct 23 Monday , 45m before sunrise
6:34 am CDT = 11:34 Universal Time
view from latitude 40°N, longitude 90°W
sidereal time 115° = 7.68h
Julian Date 2460240.98
1 day after First Quarter Moon.
westernmost elongation

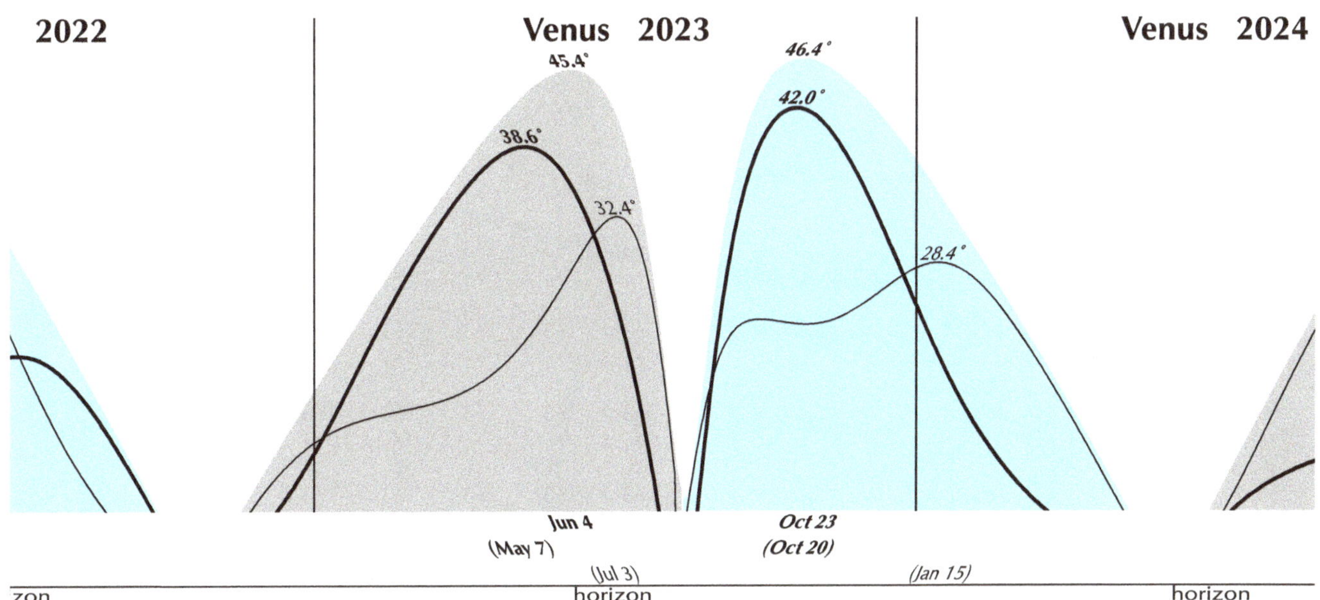
2022
Venus 2023
Venus 2024
45.4°
38.6°
32.4°
46.4°
42.0°
28.4°
Jun 4
(May 7)
(Jul 3)
Oct 23
(Oct 20)
(Jan 15)
zon
horizon
horizon

2024-2026 apparition pair, Aquarius-centered

EVENING SKY

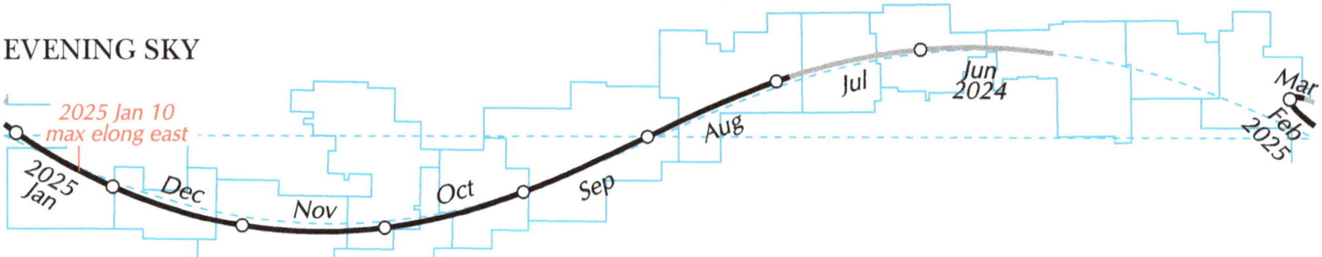

MORNING SKY

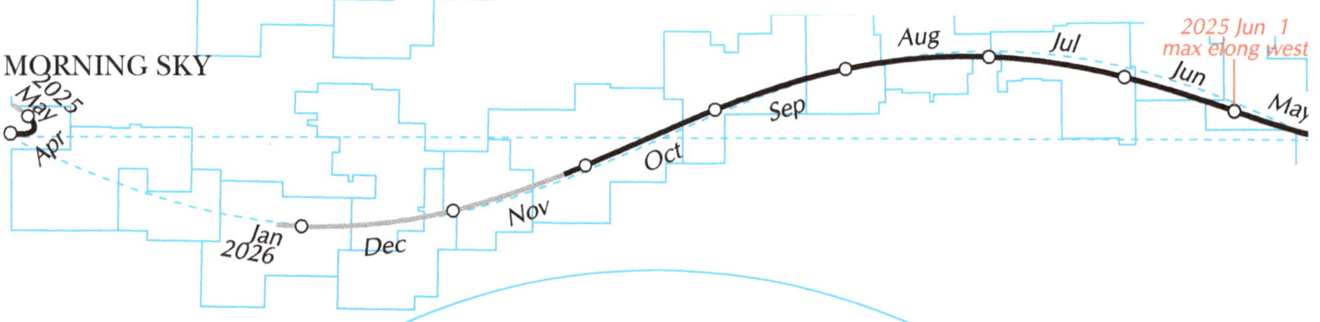

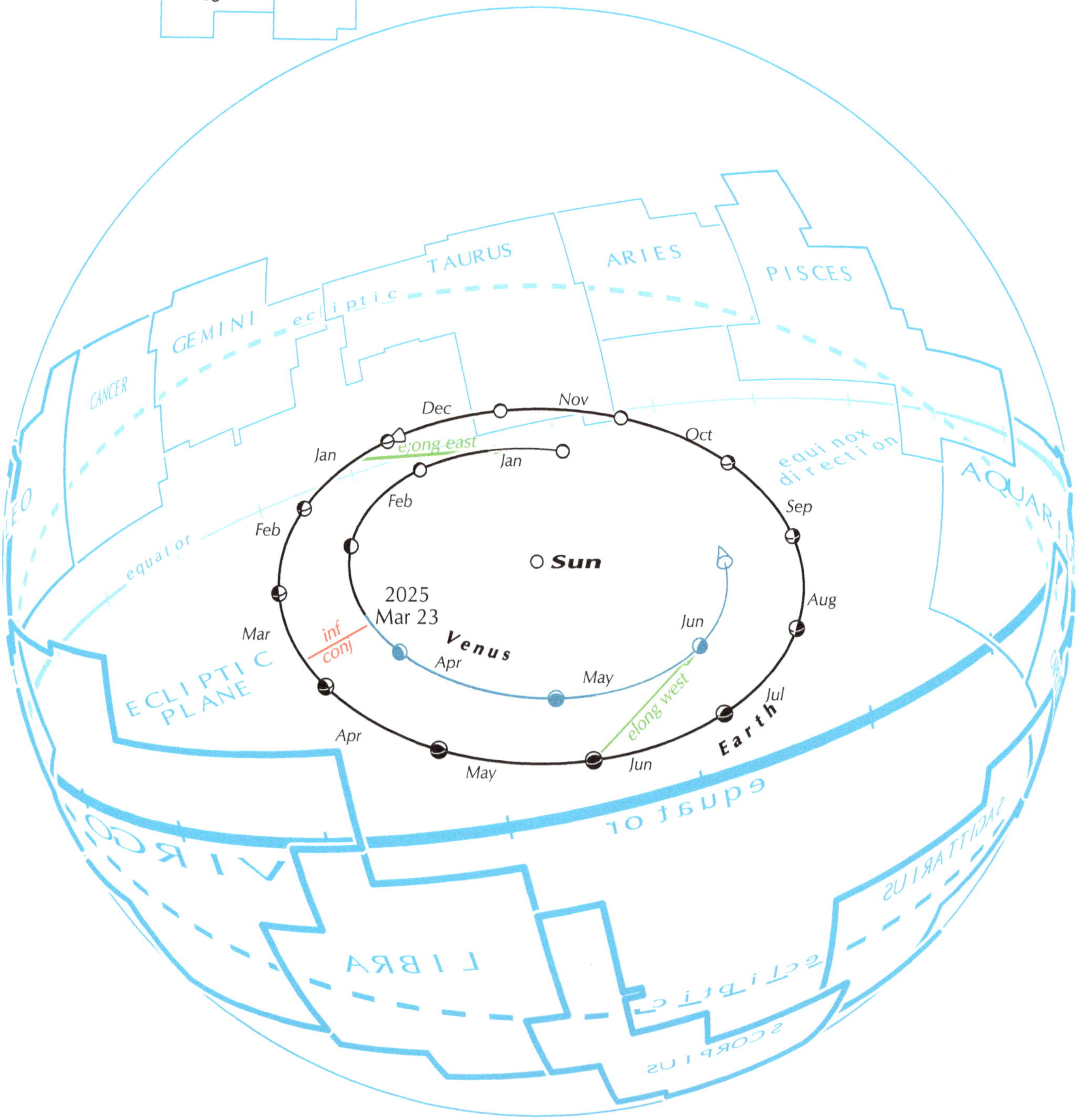

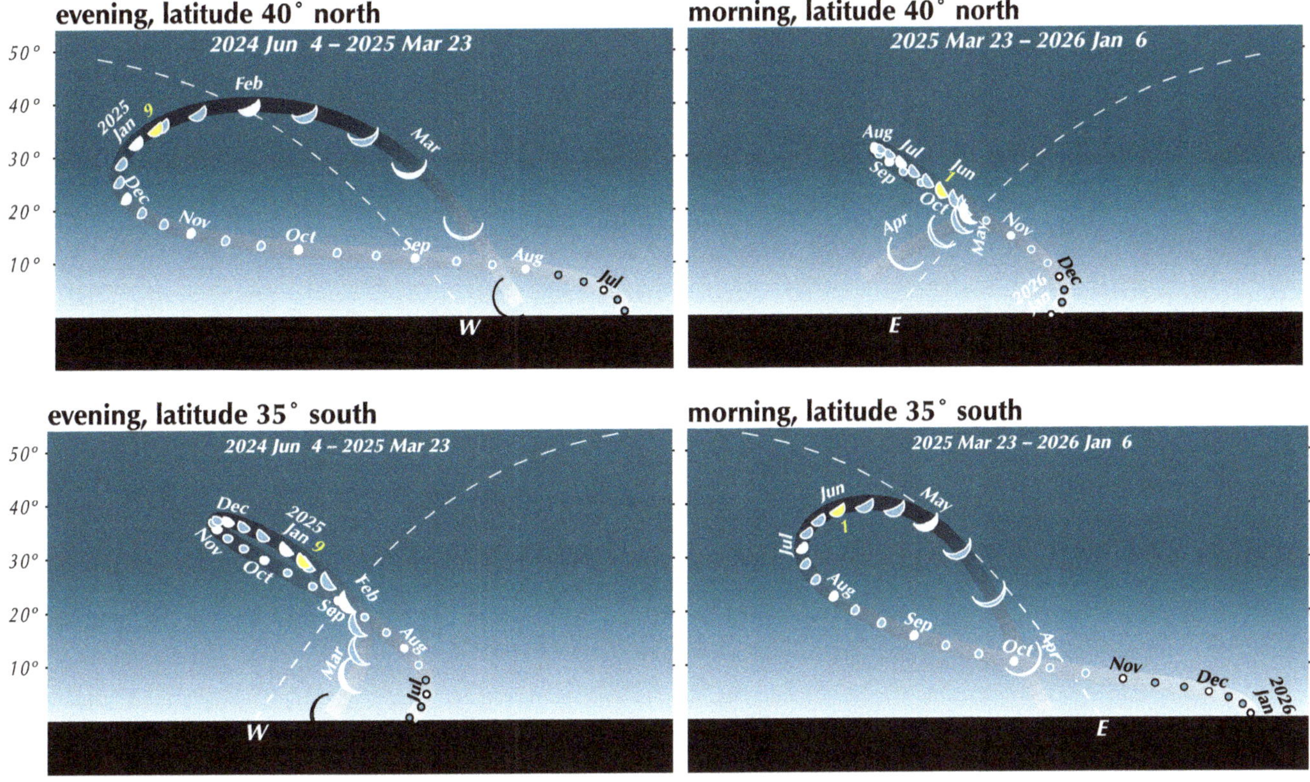

2024 Jun 4	**superior conjunction, in Taurus**
2025 Jan 10	**easternmost elongation, in Aquarius**
2025 Mar 23	**inferior conjunction, in Pisces**
2025 Jun 1	**westernmost elongation, in Pisces**
2026 Jan 6	**superior conjunction, in Sagittarius**

The course out into the evening sky is a long dip southward until November.

Then somewhat north to an easternmost elongation in Aquarius, nearly 10° south of the celestial equator, yet fairly favorable (for northerners) since the Sun, farther back along the ecliptic, is 12° more southerly.

The retrograde loop in Pisces (to and fro across the zero line of the sky) takes Venus up to the highest latitude of its orbit on Mar. 13, so that 10 days later at inferior conjunction it passes 8° north of the Sun–the widest clearance of the Sun in the whole cycle.

The morning extreme of elongation is a bit farther north in Pisces. Venus's course as it recedes through Aries, Taurus, and Gemini is south of the northerly ecliptic. Over the last five months of 2025 it travels southward, to disappear behind the Sun in Sagittarius.

2024		RA (2000)	decl	hedis	gedis	elo	mag	dia″	lum	dsu
Jun 4 15	sup.conj.with Sun	4 51 14	22 26	0.721	1.735	0	-3.9	9.6	100	0
Jun 5 10	brightest	4 55 18	22 34	0.721	1.735	0	-3.9	9.6	100	0
Jun 6 3	ascending node	4 59 9	22 42	0.721	1.735	0	-3.9	9.6	100	0
Jun 17 13	0.9°s of Mercury	5 59 59	23 53	0.719	1.731	4	-3.9	9.7	100	0
Jun 20 11	max.declin.north	6 15 52	23 57	0.719	1.728	4	-3.9	9.7	100	0
Jul 10 5	perihelion	8 0 51	21 44	0.718	1.699	10	-3.9	9.9	99	0
Jul 31 15	max.lat.north	9 48 2	14 52	0.719	1.642	16	-3.9	10.2	96	-3
Aug 6 15	5.9°N of Mercury	10 16 26	12 18	0.720	1.622	17	-3.9	10.3	95	-4
Aug 31 12	on equat.,to sou.	12 9 15	0 8	0.723	1.520	24	-3.9	11.0	91	-8
Sep 25 16	descending node	14 2 26	-12 26	0.726	1.392	30	-3.9	12.0	86	-11
Oct 30 14	aphelion	16 54 28	-24 17	0.728	1.184	38	-4.0	14.1	77	-10
Nov 14 4	max.declin.south	18 11 11	-25 37	0.728	1.089	41	-4.1	15.4	73	-7
Nov 21 13	max.lat.south	18 49 54	-25 18	0.727	1.039	42	-4.1	16.1	71	-5
Dec 7 18	0.9°N of Pluto	20 11 60	-22 25	0.726	0.927	44	-4.2	18.1	65	0

2025		RA (2000)	decl	hedis	gedis	elo	mag	dia″	lum	dsu
Jan 1 0		22 0 15	-13 41	0.723	0.751	47	-4.4	22.3	55	9
Jan 10 5	MAX.ELONG.EAS	22 35 55	-9 32	0.721	0.683	47	-4.4	24.5	51	12
Jan 12 2	dichotomy	22 42 50	-8 40	0.721	0.669	47	-4.4	25.0	50	13
Jan 16 20	ascending node	22 59 34	-6 26	0.721	0.634	47	-4.5	26.4	47	14
Jan 20 5	2.5°N of Saturn	23 10 59	-4 49	0.720	0.608	47	-4.5	27.5	45	15
Jan 30 7	on equat.,to nor.	23 41 39	0 -7	0.719	0.535	45	-4.6	31.3	39	18
Feb 3 20	4.0°N of Neptune	23 53 36	1 53	0.719	0.502	44	-4.6	33.3	36	18
Feb 14 22	max.illum.area	0 16 40	6 23	0.718	0.426	40	-4.6	39.3	27	19
Feb 16 9	brightest	0 18 52	6 54	0.718	0.417	40	-4.6	40.2	26	19
Feb 19 20	perihelion	0 23 13	8 2	0.718	0.395	38	-4.6	42.4	23	19
Feb 28 4	stat.in r.a.>retr	0 27 52	10 9	0.719	0.347	32	-4.6	48.3	15	18
Mar 2 1	stat.in long>retr	0 27 37	10 29	0.719	0.337	30	-4.6	49.6	14	18
Mar 9 11	6.3°N of Mercury	0 21 36	11 1	0.719	0.305	22	-4.4	54.8	7	15
Mar 13 8	max.lat.north	0 15 29	10 47	0.719	0.293	17	-4.3	57.1	4	14

============

2025		RA (2000)	decl	hedis	gedis	elo	mag	dia″	lum	dsu
Mar 23 1	inf.conj.with Sun	23 54 59	8 38	0.720	0.281	-8	-4.0	59.6	1	8
Apr 10 15	stat.in r.a.>dir.	23 31 12	2 33	0.723	0.330	-27	-4.4	50.7	11	-6
Apr 13 1	stat.in long>dir.	23 31 38	2 0	0.723	0.342	-29	-4.5	48.9	13	-7
Apr 24 6	brightest	23 43 50	0 45	0.724	0.410	-37	-4.5	40.8	23	-12
Apr 27 17	max.illum.area	23 50 23	0 46	0.725	0.434	-39	-4.5	38.6	26	-13
Apr 29 2	3.7°N of Saturn	23 53 19	0 49	0.725	0.444	-40	-4.5	37.7	27	-14
May 4 3	2.1°N of Neptune	0 5 9	1 13	0.726	0.480	-42	-4.5	34.9	31	-15
May 8 9	descending node	0 16 31	1 47	0.726	0.512	-43	-4.5	32.7	35	-15
Jun 1 3	MAX.ELONG.WES	1 34 46	7 34	0.728	0.700	-46	-4.3	23.9	50	-14
Jun 1 22	dichotomy	1 37 35	7 48	0.728	0.706	-46	-4.3	23.7	50	-14
Jun 12 3	aphelion	2 16 44	10 58	0.728	0.788	-45	-4.2	21.2	55	-12
Jul 4 1	2.4°s of Uranus	3 49 26	17 25	0.727	0.960	-43	-4.1	17.4	65	-5
Jul 4 6	max.lat.south	3 50 17	17 28	0.727	0.961	-43	-4.1	17.4	65	-5
Aug 5 8	max.declin.north	6 24 24	22 1	0.724	1.192	-37	-4.0	14.0	77	5
Aug 12 8	0.9°s of Jupiter	6 59 40	21 48	0.723	1.238	-36	-4.0	13.5	79	7
Aug 29 13	ascending node	8 26 18	19 12	0.720	1.343	-32	-4.0	12.5	84	10
Oct 2 9	perihelion	11 8 19	6 58	0.718	1.512	-24	-3.9	11.1	91	11
Oct 16 17	on equat.,to sou.	12 14 3	0 8	0.719	1.568	-20	-3.9	10.7	94	9
Oct 24 1	max.lat.north	12 47 32	-3 25	0.719	1.592	-18	-3.9	10.5	95	8
Nov 25 5	1.1°s of Mercury	15 21 36	-17 31	0.723	1.672	-10	-3.9	10.0	98	3
Dec 19 2	descending node	17 27 48	-23 13	0.726	1.703	-4	-3.9	9.8	100	0
Dec 27 22	max.declin.south	18 16 13	-23 43	0.727	1.709	-2	-3.9	9.8	100	0

2026		RA (2000)	decl	hedis	gedis	elo	mag	dia″	lum	dsu
Jan 1 0		18 38 41	-23 38	0.727	1.710	-1	-3.9	9.8	100	-1

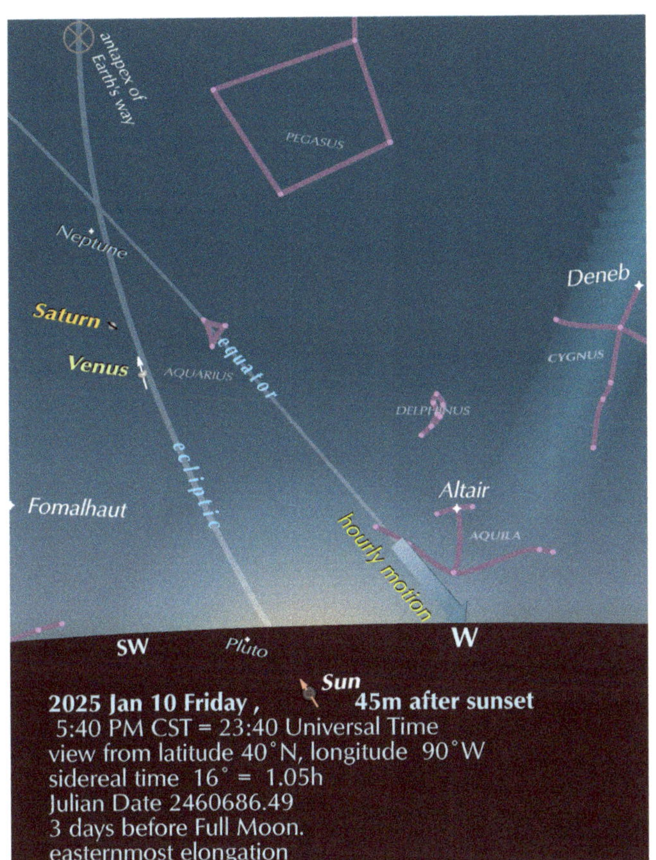
antapex of Earth's way
PEGASUS
Neptune
Saturn
Venus
AQUARIUS
equator
ecliptic
Deneb
CYGNUS
DELPHINUS
Fomalhaut
Altair
AQUILA
hourly motion
SW
Pluto
W
Sun
2025 Jan 10 Friday , 45m after sunset
5:40 PM CST = 23:40 Universal Time
view from latitude 40°N, longitude 90°W
sidereal time 16° = 1.05h
Julian Date 2460686.49
3 days before Full Moon.
easternmost elongation

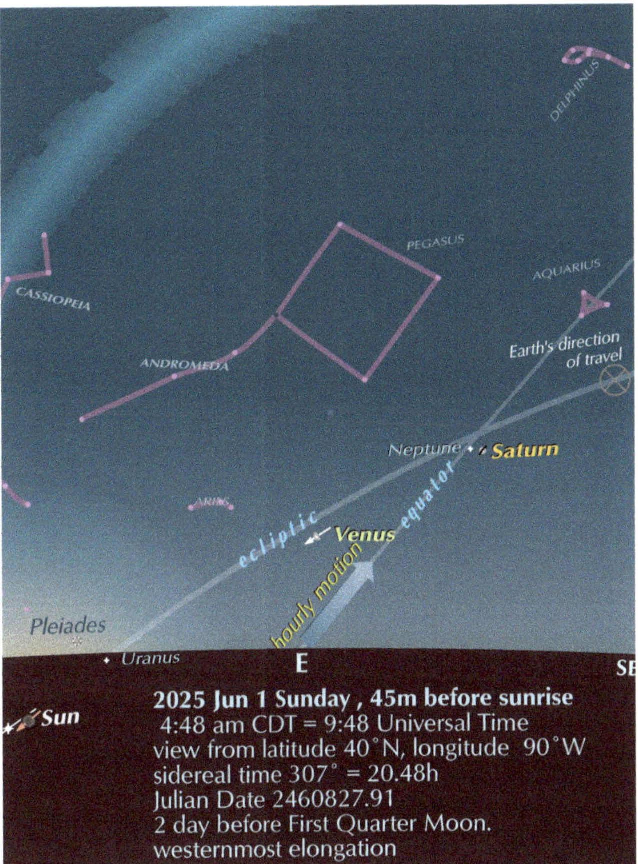
DELPHINUS
PEGASUS
AQUARIUS
CASSIOPEIA
ANDROMEDA
Earth's direction of travel
Neptune
Saturn
ecliptic
Venus
equator
hourly motion
Pleiades
Uranus
E
Sun
2025 Jun 1 Sunday , 45m before sunrise
4:48 am CDT = 9:48 Universal Time
view from latitude 40°N, longitude 90°W
sidereal time 307° = 20.48h
Julian Date 2460827.91
2 day before First Quarter Moon.
westernmost elongation

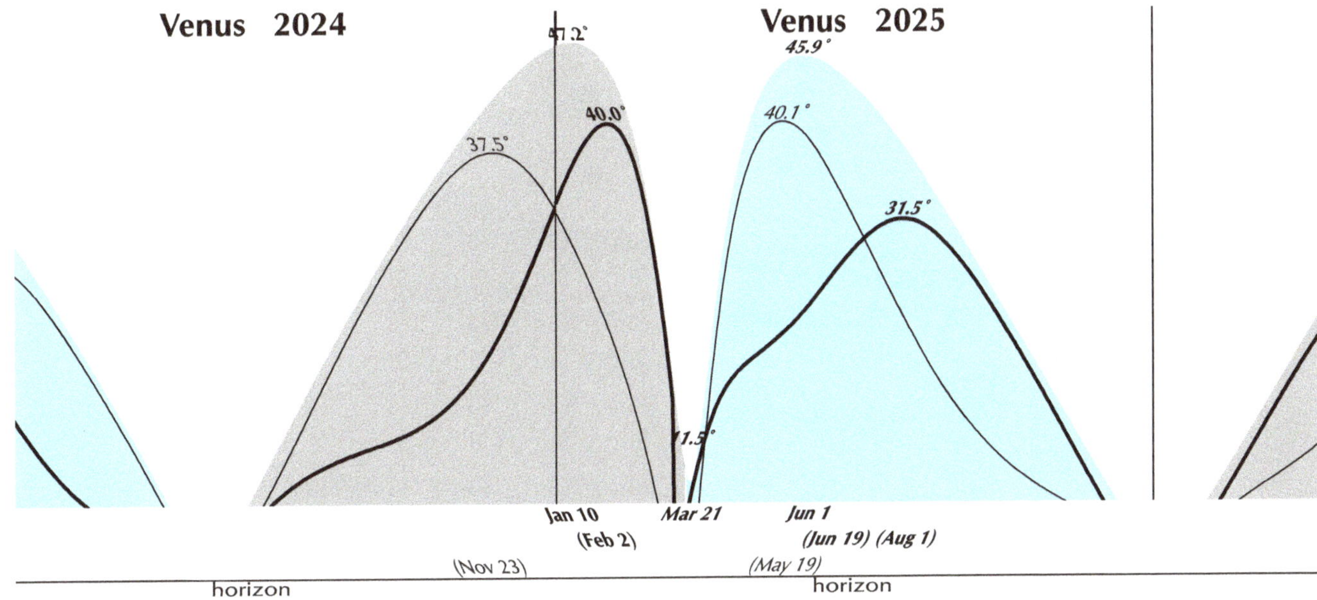
Venus 2024
Venus 2025
45.9°
40.0°
40.1°
37.5°
31.5°
11.5°
Jan 10
(Feb 2)
Mar 21
Jun 1
(Jun 19) (Aug 1)
(Nov 23)
(May 19)
horizon
horizon

2026-2027 apparition pair, Virgo-centered

EVENING SKY

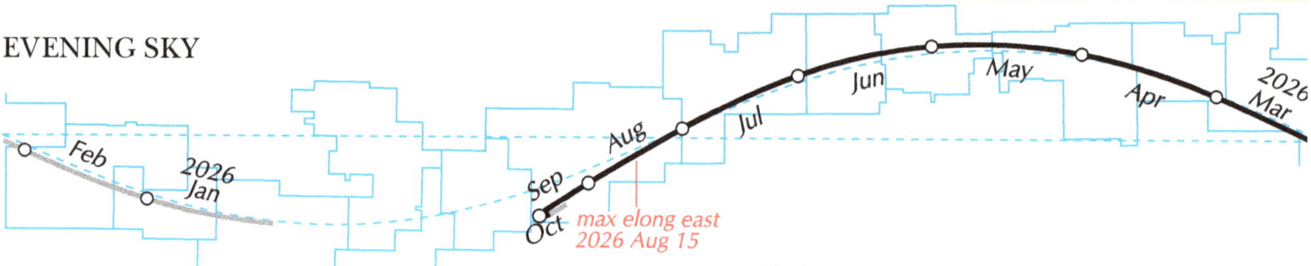

2026 Mar Apr May Jun Jul Aug Sep Oct

max elong east
2026 Aug 15

Feb 2026 Jan

MORNING SKY

Jun May Jul Aug

Nov Oct 2026 Dec

2027 Jan 3
max elong west

Jan 2027 Feb Mar Apr

TAURUS ARIES PISCES GEMINI CANCER ecliptic LEO AQUARIUS

equinox direction

equator

ECLIPTIC PLANE

inf conj
2026
Oct 24

elong west
elong east

Sun

Venus

Earth

VIRGO LIBRA SCORPIUS SAGITTARIUS

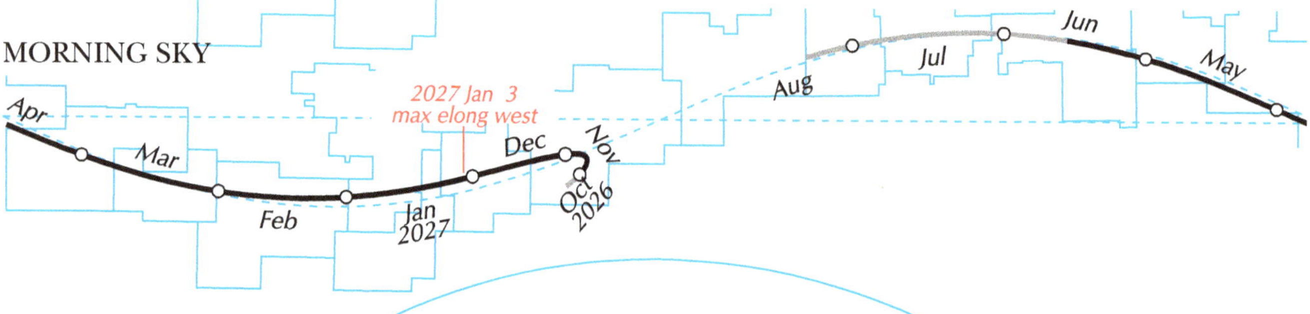

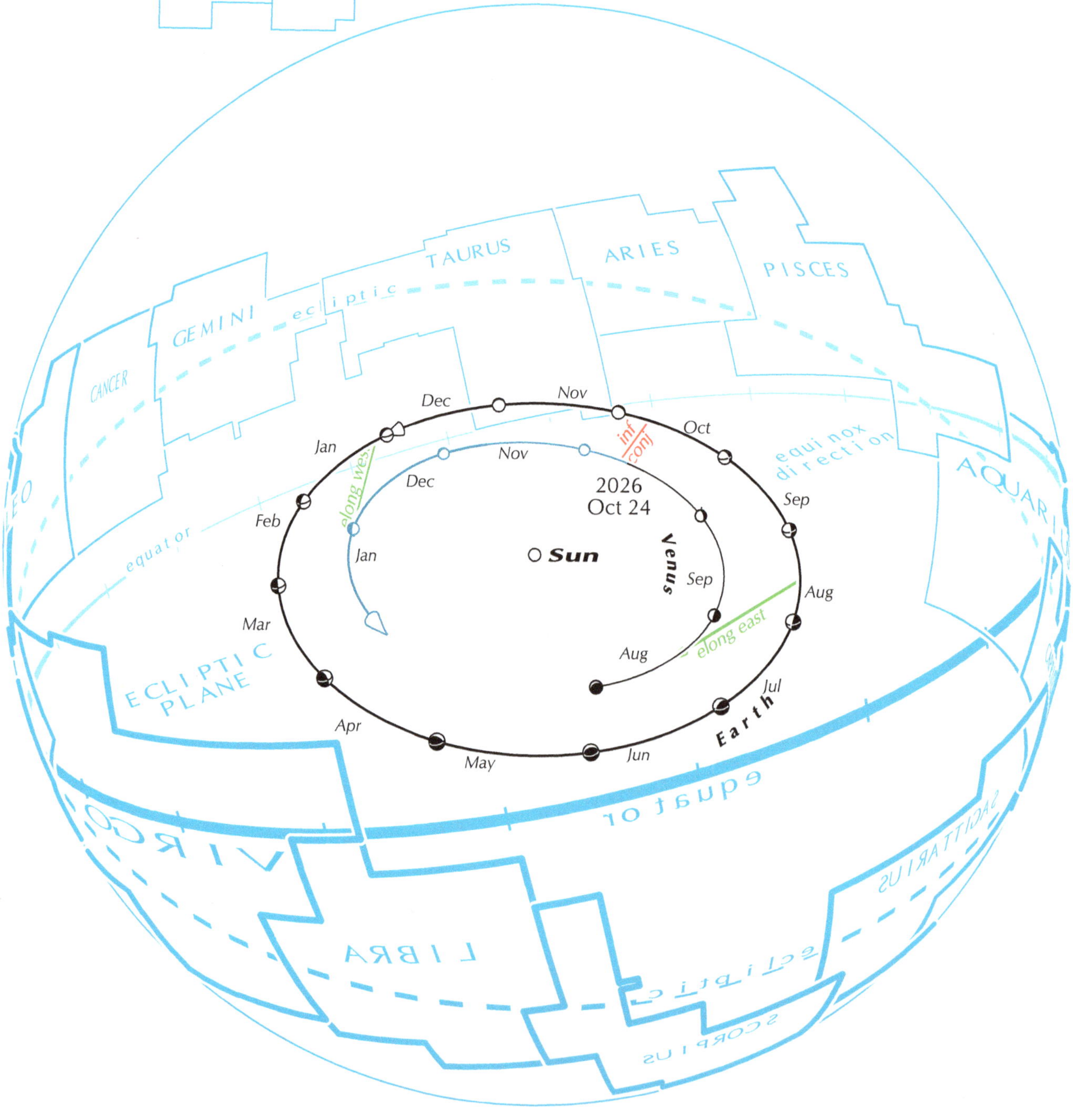

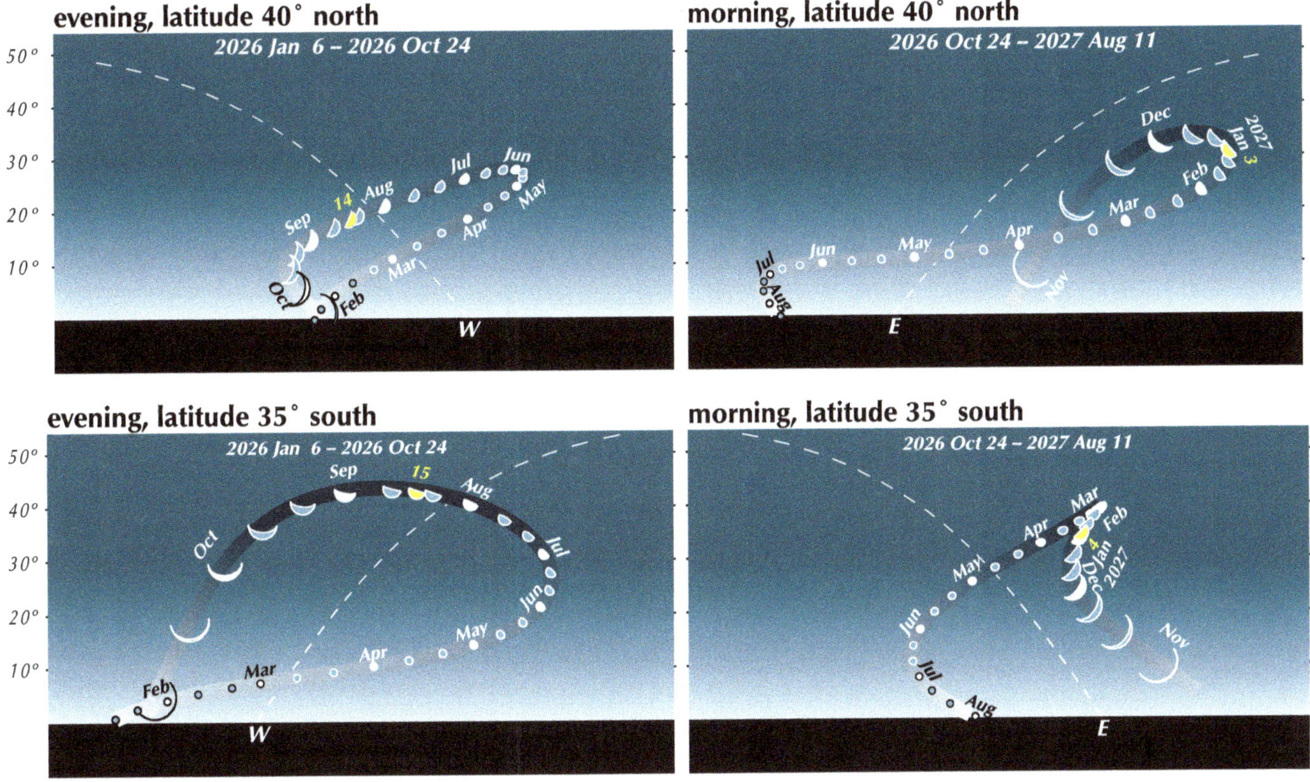

2026 Jan 6	**superior conjunction, in Sagittarius**
2026 Aug 15	**easternmost elongation, in Virgo**
2026 Oct 24	**inferior conjunction, in Virgo**
2027 Jan 3	**westernmost elongation, in Libra**
2027 Aug 11	**superior conjunction, in Cancer**

From behind the Sun, Venus moves out northward till May, then southward till August, and is slightly south of the equator by the time of easternmost elongation. This gives the more favorable show for the southern hemisphere.

The retrograde motion after that is, on the map, a narrow loop traced in the southernmost extension of Virgo—a region that we might call the trailing robe of the goddess. So at inferior conjunction Venus passes 6° south of the Sun—the second greatest such southern clearance in the cycle.

Out in the morning sky, Venus is at greatest elongation quite far south, but 7° farther north than the Sun, so that (for the north hemisphere) it is quite high in the dawn skies.

2026		RA (2000)	decl	hedis	gedis	elo	mag	dia″	lum	dsu
Jan 6 16	sup.conj.with Sun	19 9 43	-23 10	0.728	1.711	1	-3.9	9.8	100	-1
Jan 8 4	0.2°N of Mars	19 17 50	-22 59	0.728	1.711	1	-3.9	9.8	100	-1
Jan 20 17	2.7°N of Pluto	20 24 55	-20 25	0.728	1.708	4	-3.9	9.8	100	0
Jan 22 19	aphelion	20 35 49	-19 50	0.728	1.707	4	-3.9	9.8	100	0
Jan 28 24	0.7°N of Mercury	21 7 37	-17 51	0.728	1.702	5	-3.9	9.8	100	0
Feb 13 22	max.lat.south	22 25 52	-11 23	0.727	1.684	9	-3.9	9.9	99	2
Feb 26 23	4.7°s of Mercury	23 26 35	-5 6	0.726	1.661	12	-3.9	10.1	98	3
Mar 7 12	0.1°N of Neptune	0 5 28	0-44	0.725	1.642	14	-3.9	10.2	97	5
Mar 8 17	on equat.,to nor.	0 10 51	0 -8	0.725	1.640	15	-3.9	10.2	97	5
Mar 8 22	1.0°N of Saturn	0 11 54	0 0	0.725	1.639	15	-3.9	10.2	97	5
Apr 11 5	ascending node	2 45 33	15 59	0.720	1.530	23	-3.9	10.9	92	8
Apr 24 5	0.8°N of Uranus	3 49 31	20 40	0.719	1.472	26	-3.9	11.4	89	8
May 15 3	perihelion	5 37 42	24 48	0.718	1.361	31	-4.0	12.3	84	6
May 21 18	max.declin.north	6 12 36	25 5	0.719	1.321	33	-4.0	12.7	83	5
Jun 5 17	max.lat.north	7 30 27	23 50	0.719	1.225	36	-4.0	13.7	78	1
Jun 9 12	1.6°N of Jupiter	7 49 35	23 7	0.720	1.199	37	-4.0	14.0	77	0
Jul 31 19	descending node	11 36 0	2 36	0.726	0.802	45	-4.2	20.9	56	-16
Aug 5 19	on equat.,to sou.	11 54 10	0 9	0.727	0.762	46	-4.3	22.0	54	-17
Aug 12 20	dichotomy	12 18 47	-3 16	0.727	0.706	46	-4.3	23.7	50	-18
Aug 15 6	MAX.ELONG.EAS	12 27 3	-4 26	0.727	0.686	46	-4.3	24.4	49	-19
Sep 4 12	aphelion	13 28 50	-13 25	0.728	0.528	44	-4.5	31.7	36	-21
Sep 18 23	max.illum.area	14 0 55	-18 23	0.728	0.423	39	-4.6	39.6	26	-20
Sep 22 18	brightest	14 6 31	-19 22	0.728	0.398	37	-4.6	42.1	22	-20
Sep 26 15	max.lat.south	14 10 35	-20 11	0.727	0.373	35	-4.6	44.9	19	-19
Oct 2 14	stat.in r.a.>retr	14 13 3	-21 0	0.727	0.339	30	-4.5	49.4	14	-17
Oct 3 7	stat.in long>retr	14 13 1	-21 3	0.727	0.335	29	-4.5	49.9	13	-17
Oct 5 15	5.4°s of Mercury	14 12 22	-21 9	0.727	0.324	27	-4.5	51.7	11	-16
Oct 6 18	max.declin.south	14 11 47	-21 10	0.726	0.318	26	-4.5	52.6	10	-16.6

============

2026		RA (2000)	decl	hedis	gedis	elo	mag	dia″	lum	dsu
Oct 24 4	inf.conj.with Sun	13 43 33	-17 41	0.724	0.273	-7	-4.0	61.3	1	-6
Nov 11 15	stat.in r.a.>dir.	13 20 18	-10 51	0.722	0.315	-26	-4.5	53.2	10	7
Nov 14 0	stat.in long>dir.	13 20 43	-10 17	0.721	0.326	-29	-4.6	51.4	12	8
Nov 21 22	ascending node	13 27 44	-9 11	0.721	0.369	-36	-4.6	45.4	20	11
Nov 27 19	brightest	13 37 46	-9 5	0.720	0.405	-39	-4.7	41.3	25	12
Nov 29 14	max.illum.area	13 41 34	-9 10	0.720	0.417	-40	-4.6	40.1	27	12
Dec 25 21	perihelion	15 2 33	-13 46	0.718	0.607	-47	-4.5	27.6	45	9

2027		RA (2000)	decl	hedis	gedis	elo	mag	dia″	lum	dsu
Jan 1 0		15 26 24	-15 15	0.719	0.653	-47	-4.5	25.6	49	8
Jan 3 8	dichotomy	15 35 55	-15 49	0.719	0.671	-47	-4.4	24.9	50	7
Jan 3 18	MAX.ELONG.WES	15 37 32	-15 55	0.719	0.674	-47	-4.4	24.8	50	7
Jan 16 10	max.lat.north	16 32 23	-18 41	0.719	0.770	-46	-4.3	21.7	56	2
Feb 9 17	max.declin.south	18 28 34	-21 15	0.722	0.951	-44	-4.2	17.6	66	-7
Mar 7 16	4.9°N of Pluto	20 38 4	-18 7	0.725	1.132	-39	-4.1	14.8	75	-13
Mar 13 12	descending node	21 6 35	-16 33	0.726	1.171	-38	-4.0	14.3	77	-14
Apr 17 5	aphelion	23 47 50	-2 54	0.728	1.383	-31	-3.9	12.1	86	-13
Apr 23 4	on equat.,to nor.	0 14 27	0 -8	0.728	1.416	-29	-3.9	11.8	87	-12
Apr 24 9	0.3°s of Neptune	0 19 52	0 25	0.728	1.422	-29	-3.9	11.8	87	-12
May 8 0	0.7°N of Saturn	1 21 8	6 46	0.727	1.492	-26	-3.9	11.2	90	-10
May 9 8	max.lat.south	1 27 9	7 22	0.727	1.499	-25	-3.9	11.2	90	-10
Jun 13 24	0.7°s of Uranus	4 18 18	20 33	0.723	1.642	-16	-3.9	10.2	96	-3
Jul 1 10	4.6°N of Mercury	5 49 34	23 17	0.721	1.689	-11	-3.9	9.9	98	0
Jul 4 15	ascending node	6 6 35	23 26	0.721	1.696	-11	-3.9	9.9	98	1
Jul 6 15	max.declin.north	6 17 21	23 27	0.720	1.700	-10	-3.9	9.8	98	1
Aug 7 14	perihelion	9 4 16	17 57	0.718	1.732	-2	-3.9	9.7	100	1
Aug 11 2	0.5°s of Mercury	9 21 37	16 44	0.718	1.732	1	-3.9	9.7	100	1

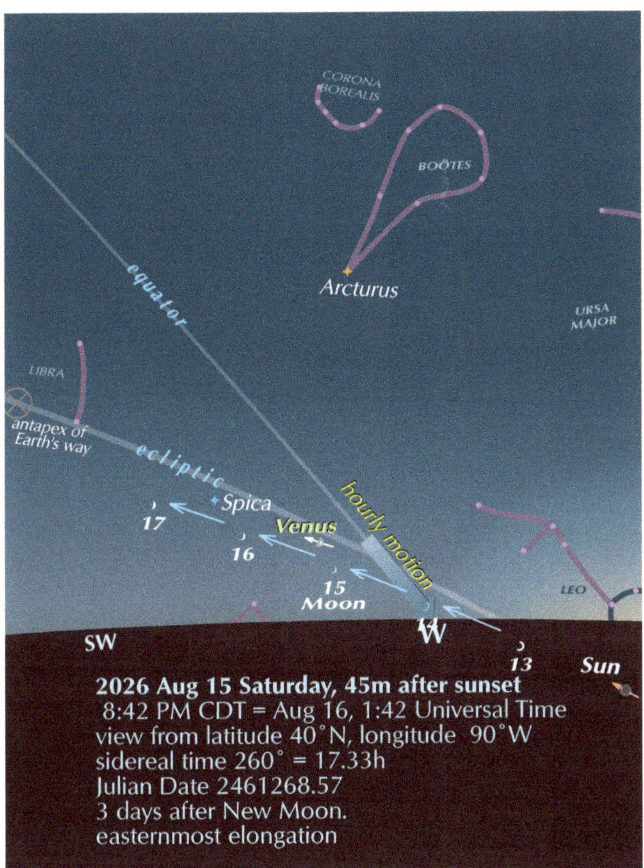

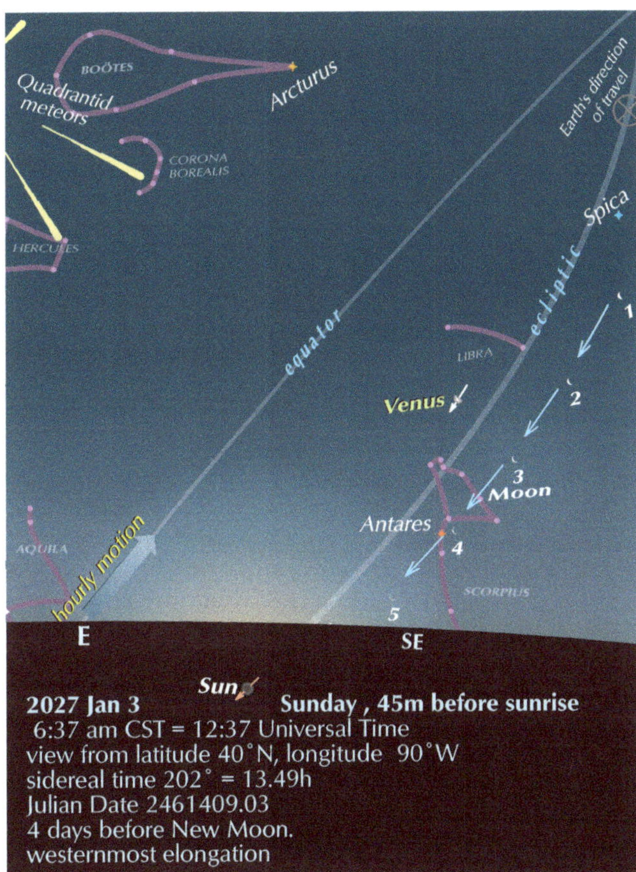

January 3 is the usual date of the annual Quadrantid meteors. The shower was so named because its radiant is in a now obsolete constellation called Quadrans Muralis, the "wall quadrant", near the end of the Big Dipper's handle.

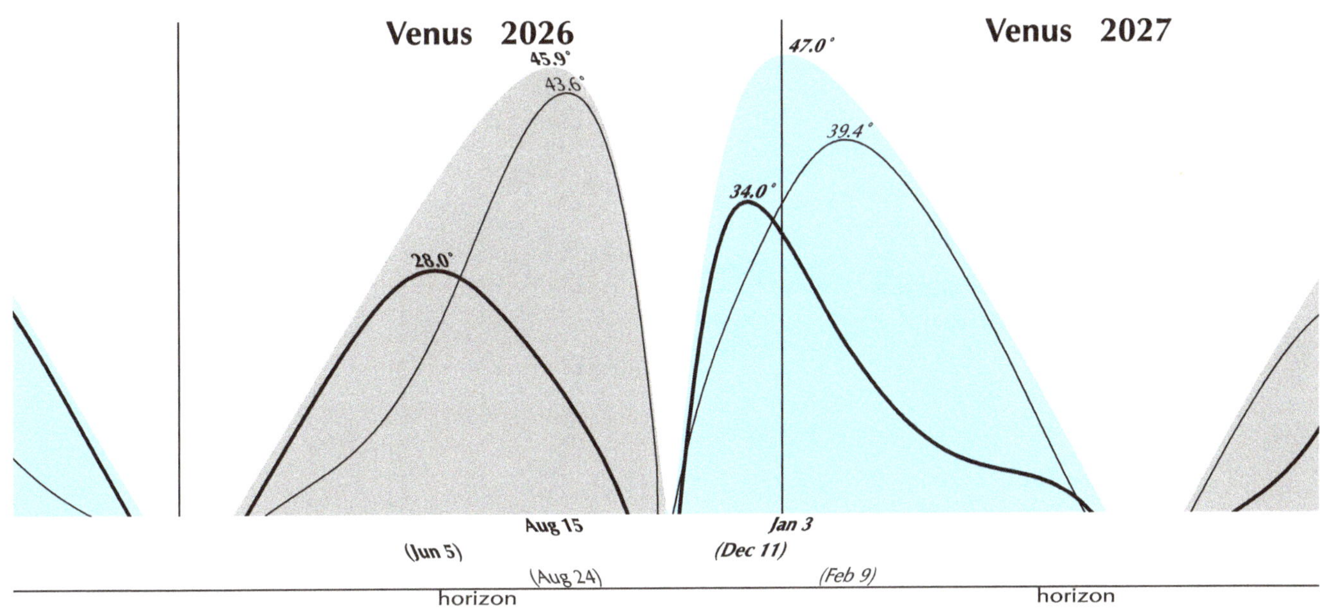

2027-2029 apparition pair, Taurus-centered

EVENING SKY

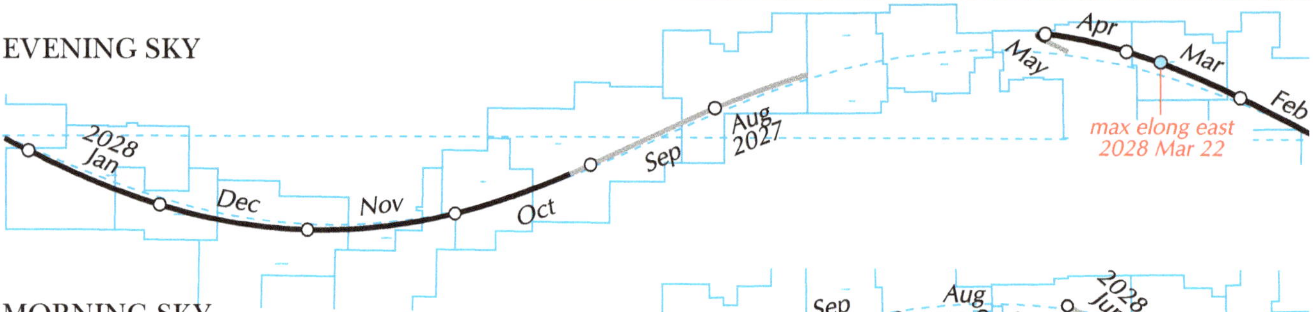

MORNING SKY

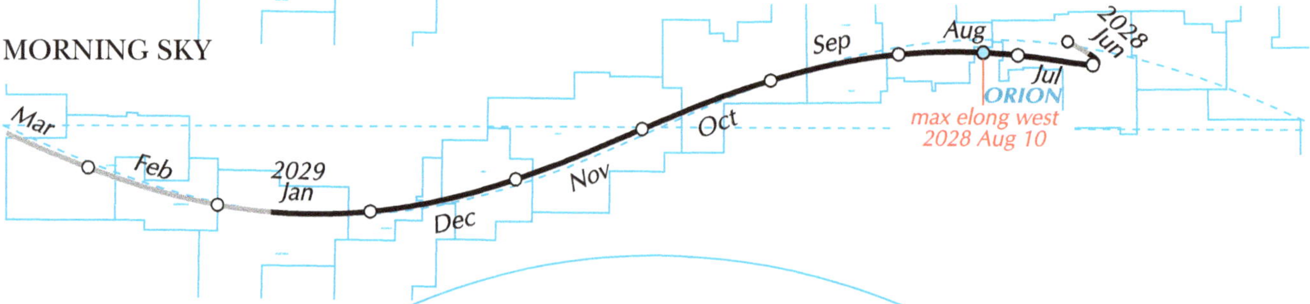

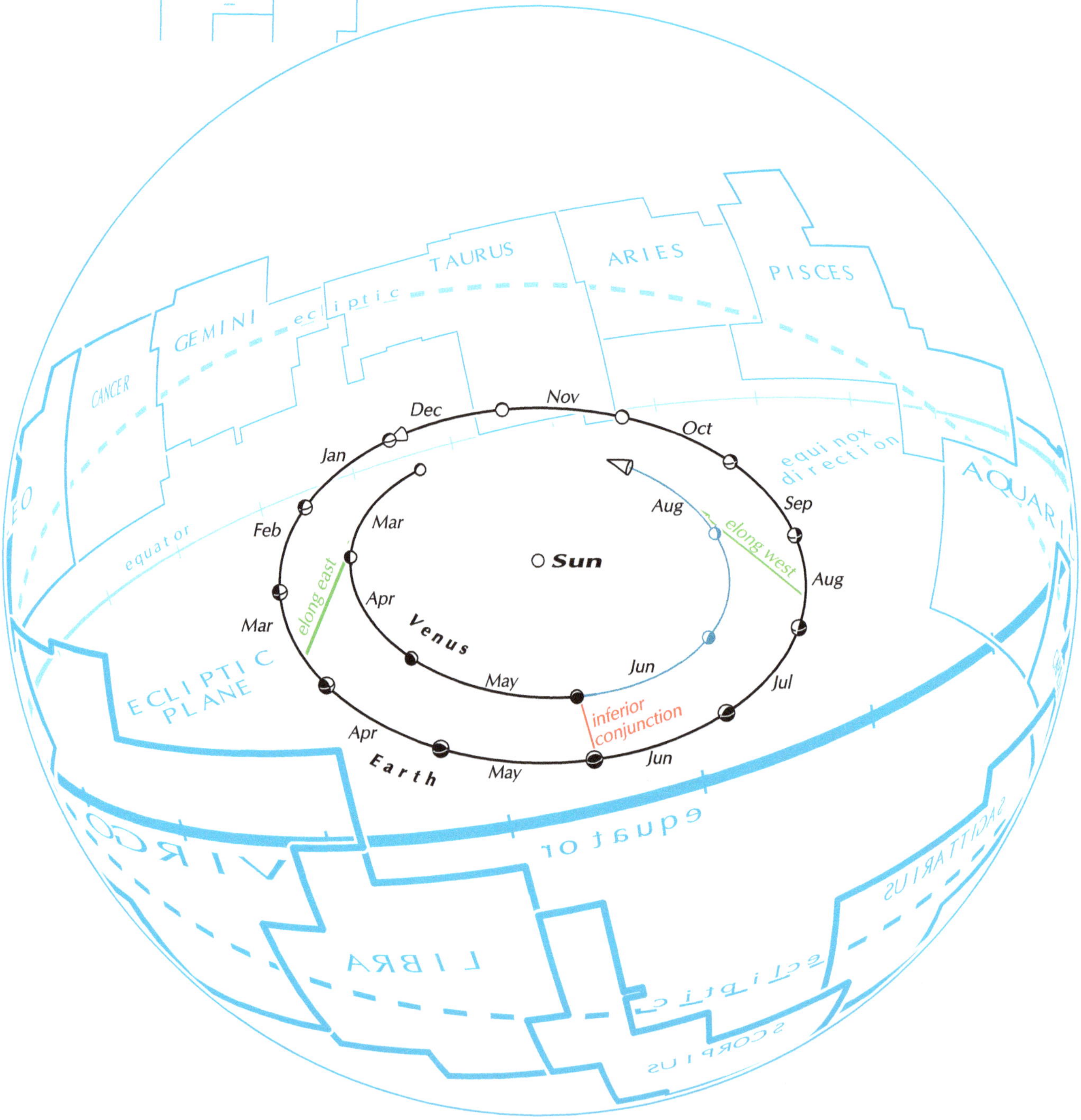

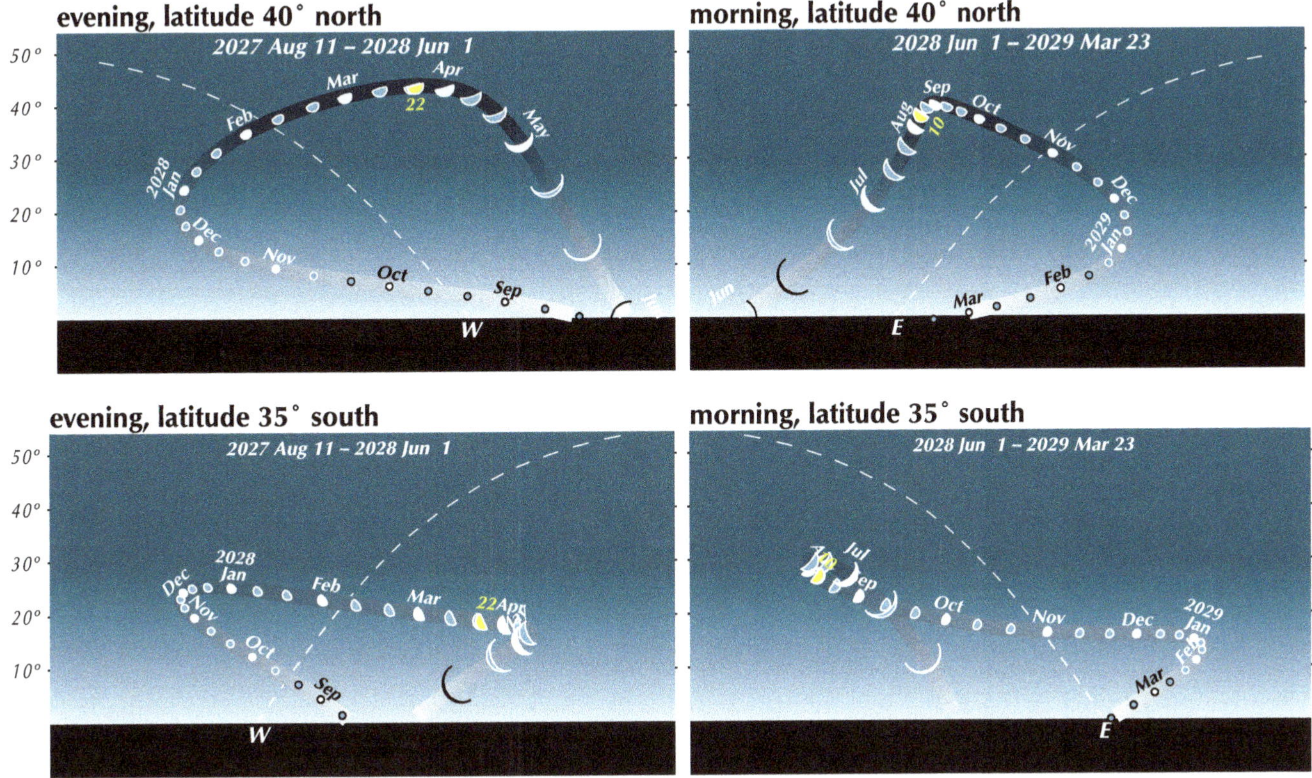

We consider this sixth apparition pair in order to show its close similarity to the first. The dates of the events are from 1 to 3 calendar days earlier, and their positions have moved backward–westward–by 1.5° (on average, and in relation to the orbital plane).

As in the earlier incarnation, the main events take place along the northern arch of the ecliptic, a fine zodiacal stage for north-hemisphere observers, though a mite less so than in 2020. The evening extreme of elongation is again in Aries; the morning one has moved across the border from Gemini into the northern tip of the "Club" of Orion.

2028 is again, like 2020, what we might call a "transitoid" year, in which inferior conjunction falls close to descending node in Taurus, so that Venus passes quite close to the Sun rather than widely north or south of it. But the clearance is another 0.3° farther north, so for a second time since 2012 missing a transit.

Venus again plows through the Pleiades, more deeply. And again reaches the northernmost declination of the cycle, though having moved back westward it is not so near to the star El Nath on the Taurus-Auriga border, and is 3′ farther south.

Odi et amo. Quare id faciam fortasse requiris.
Nescio, sed fieri sentio et excrucior.

I hate and love. Why so, you well may ask?
I know not, but I feel it and am tortred.

Catullus, *Carmina*, 85, referring to the woman he called Lesbia. She was cool toward him, had many other admirers, and was probably Clodia "Pulchra" ("beautiful"), a married aristocrat.

2027		RA (2000)	decl	hedis	gedis	elo	mag	dia"	lum	dsu
Aug 11 24	sup.conj.with Sun	9 26 11	16 24	0.719	1.731	1	-3.9	9.7	100	1
Aug 13 11	brightest	9 33 17	15 52	0.719	1.731	1	-3.9	9.7	100	1
Aug 25 22	0.5°N of Jupiter	10 33 3	10 38	0.719	1.724	4	-3.9	9.7	100	0
Aug 29 3	max.lat.north	10 47 59	9 10	0.719	1.720	5	-3.9	9.7	100	0
Sep 16 8	on equat.,to sou.	12 11 17	0 9	0.721	1.692	10	-3.9	9.9	99	-3
Oct 10 13	4.2°N of Mercury	14 2 19	-11 50	0.724	1.630	16	-3.9	10.3	96	-5
Oct 24 4	descending node	15 8 18	-17 36	0.726	1.585	19	-3.9	10.6	95	-6
Nov 25 2	0.3°s of Mars	17 56 1	-24 44	0.728	1.454	27	-3.9	11.5	89	-4
Nov 27 20	aphelion	18 11 10	-24 48	0.728	1.441	28	-3.9	11.6	89	-4
Nov 28 4	max.declin.south	18 12 60	-24 48	0.728	1.440	28	-3.9	11.6	89	-4
Dec 20 1	max.lat.south	20 10 8	-21 57	0.727	1.330	32	-4.0	12.6	84	2
Dec 25 2	3.0°N of Pluto	20 36 6	-20 32	0.727	1.302	34	-4.0	12.9	83	3

2028		RA (2000)	decl	hedis	gedis	elo	mag	dia"	lum	dsu
Jan 1 0		21 10 36	-18 11	0.726	1.264	35	-4.0	13.2	82	5
Feb 8 5	on equat.,to nor.	0 2 40	0 -8	0.721	1.025	42	-4.1	16.3	70	15
Feb 11 24	1.3°N of Neptune	0 18 23	1 49	0.721	0.999	43	-4.2	16.8	69	16
Feb 14 8	ascending node	0 28 5	3 2	0.721	0.983	43	-4.2	17.0	68	16
Mar 1 12	3.8°N of Saturn	1 34 23	11 11	0.719	0.867	45	-4.2	19.3	61	19
Mar 19 7	perihelion	2 46 3	18 53	0.718	0.733	46	-4.3	22.8	53	19
Mar 22 12	MAX.ELONG.EAS	2 58 48	20 4	0.718	0.709	46	-4.4	23.6	51	19
Mar 24 17	dichotomy	3 7 29	20 51	0.719	0.691	46	-4.4	24.2	50	19
Apr 9 20	max.lat.north	4 7 10	25 17	0.719	0.567	45	-4.5	29.5	40	17
Apr 14 0	4.7°N of Uranus	4 21 0	26 5	0.720	0.536	44	-4.5	31.3	37	17
Apr 25 18	max.illum.area	4 53 19	27 30	0.721	0.449	40	-4.5	37.2	28	14
Apr 26 6	brightest	4 54 28	27 32	0.721	0.446	40	-4.5	37.6	27	14
May 3 2	max.declin.north	5 6 25	27 45	0.722	0.400	36	-4.5	41.9	21	12
May 10 23	stat.in long>retr	5 11 49	27 27	0.723	0.353	29	-4.4	47.5	13	10
May 11 3	stat.in r.a.>retr	5 11 49	27 26	0.723	0.352	29	-4.4	47.6	13	9
May 24 2	2.5°N of Mercury	4 57 30	25 19	0.725	0.299	13	-4.1	56.0	3	5

============

2028		RA (2000)	decl	hedis	gedis	elo	mag	dia"	lum	dsu
Jun 1 10	inf.conj.with Sun	4 37 25	22 54	0.726	0.288	-1	-2.7	58.0	0	1
Jun 3 6	0.4°N of Uranus	4 32 47	22 19	0.726	0.289	-3	-3.8	57.9	0	0
Jun 4 21	descending node	4 28 45	21 47	0.726	0.290	-5	-3.9	57.7	0	-1
Jun 14 16	2.2°s of Mars	4 10 15	18 54	0.727	0.315	-20	-4.2	53.2	6	-4
Jun 22 8	stat.in r.a.>dir.	4 5 38	17 30	0.728	0.350	-28	-4.4	47.8	13	-6
Jun 22 22	stat.in long>dir.	4 5 39	17 26	0.728	0.353	-29	-4.4	47.4	13	-6
Jul 6 4	brightest	4 18 55	17 2	0.728	0.437	-39	-4.5	38.3	25	-6
Jul 8 0	max.illum.area	4 22 26	17 7	0.728	0.450	-40	-4.5	37.2	27	-5
Jul 9 16	aphelion	4 25 51	17 13	0.728	0.462	-40	-4.5	36.3	28	-5
Jul 16 14	4.4°s of Uranus	4 42 56	17 48	0.728	0.513	-43	-4.5	32.6	34	-3
Jul 31 17	max.lat.south	5 31 59	19 21	0.727	0.630	-45	-4.4	26.6	44	1
Aug 10 13	dichotomy	6 10 1	20 3	0.726	0.707	-46	-4.3	23.7	50	5
Aug 10 16	MAX.ELONG.WES	6 10 27	20 4	0.726	0.708	-46	-4.3	23.7	50	5
Sep 8 22	2.3°s of Mars	8 19 27	18 20	0.723	0.932	-44	-4.2	18.0	64	13
Sep 26 0	ascending node	9 38 10	14 7	0.721	1.055	-41	-4.1	15.9	71	15
Oct 30 2	perihelion	12 12 19	0 23	0.718	1.275	-34	-4.0	13.1	82	14
Oct 30 14	on equat.,to sou.	12 14 30	0 9	0.718	1.277	-34	-4.0	13.1	82	14
Nov 9 19	0.7°N of Jupiter	13 0 42	-4 34	0.719	1.335	-32	-4.0	12.5	84	12
Nov 20 13	max.lat.north	13 50 15	-9 27	0.719	1.392	-29	-4.0	12.0	87	10
Jan 1 0		17 19 38	-22 29	0.724	1.569	-20	-3.9	10.7	94	0

2029		RA (2000)	decl	hedis	gedis	elo	mag	dia"	lum	dsu
Jan 1 0		17 19 38	-22 29	0.724	1.569	-20	-3.9	10.7	94	0
Jan 11 17	max.declin.south	18 17 52	-23 11	0.726	1.604	-17	-3.9	10.4	96	-2
Jan 15 14	descending node	18 38 58	-23 7	0.726	1.615	-16	-3.9	10.4	96	-2
Jan 22 24	3.7°s of Mercury	19 19 16	-22 29	0.727	1.636	-15	-3.9	10.2	97	-3
Feb 9 1	4.5°N of Pluto	20 49 13	-18 43	0.728	1.675	-11	-3.9	10.0	98	-4
Feb 19 10	aphelion	21 41 26	-15 6	0.728	1.693	-8	-3.9	9.9	99	-4
Mar 13 10	max.lat.south	23 25 55	-5 13	0.727	1.718	-3	-3.9	9.7	100	-2
Mar 23 14	on equat.,to nor.	0 12 21	0 -9	0.726	1.723	1	-3.9	9.7	100	-1
Mar 23 20	sup.conj.with Sun	0 13 24	0 -2	0.726	1.723	1	-3.9	9.7	100	-1

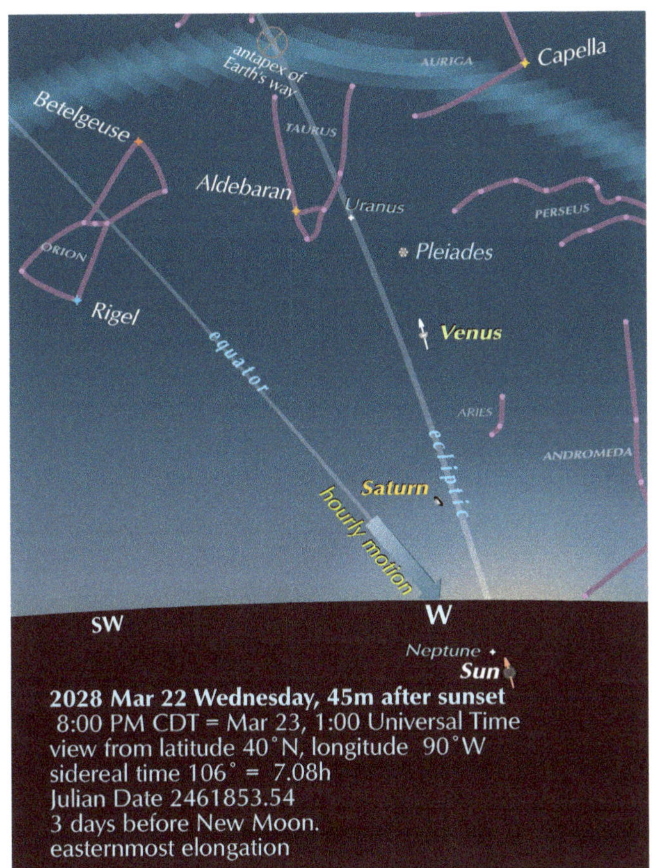
Capella
AURIGA
antapex of Earth's way
Betelgeuse
TAURUS
Aldebaran
Uranus
PERSEUS
ORION
Pleiades
Rigel
Venus
equator
ARIES
ecliptic
ANDROMEDA
Saturn
hourly motion
SW
W
Neptune
Sun
2028 Mar 22 Wednesday, 45m after sunset
8:00 PM CDT = Mar 23, 1:00 Universal Time
view from latitude 40°N, longitude 90°W
sidereal time 106° = 7.08h
Julian Date 2461853.54
3 days before New Moon.
easternmost elongation

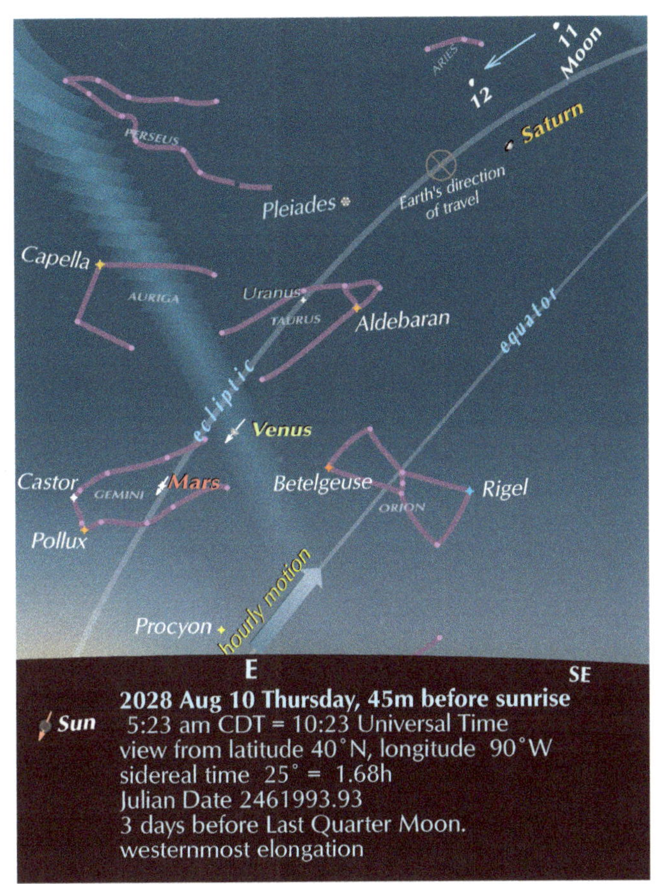
ARIES
11
Moon
12
Saturn
PERSEUS
Earth's direction of travel
Pleiades
Capella
AURIGA
Uranus
TAURUS
Aldebaran
equator
ecliptic
Venus
Castor
GEMINI
Mars
Betelgeuse
ORION
Rigel
Pollux
hourly motion
Procyon
E
SE
Sun
2028 Aug 10 Thursday, 45m before sunrise
5:23 am CDT = 10:23 Universal Time
view from latitude 40°N, longitude 90°W
sidereal time 25° = 1.68h
Julian Date 2461993.93
3 days before Last Quarter Moon.
westernmost elongation

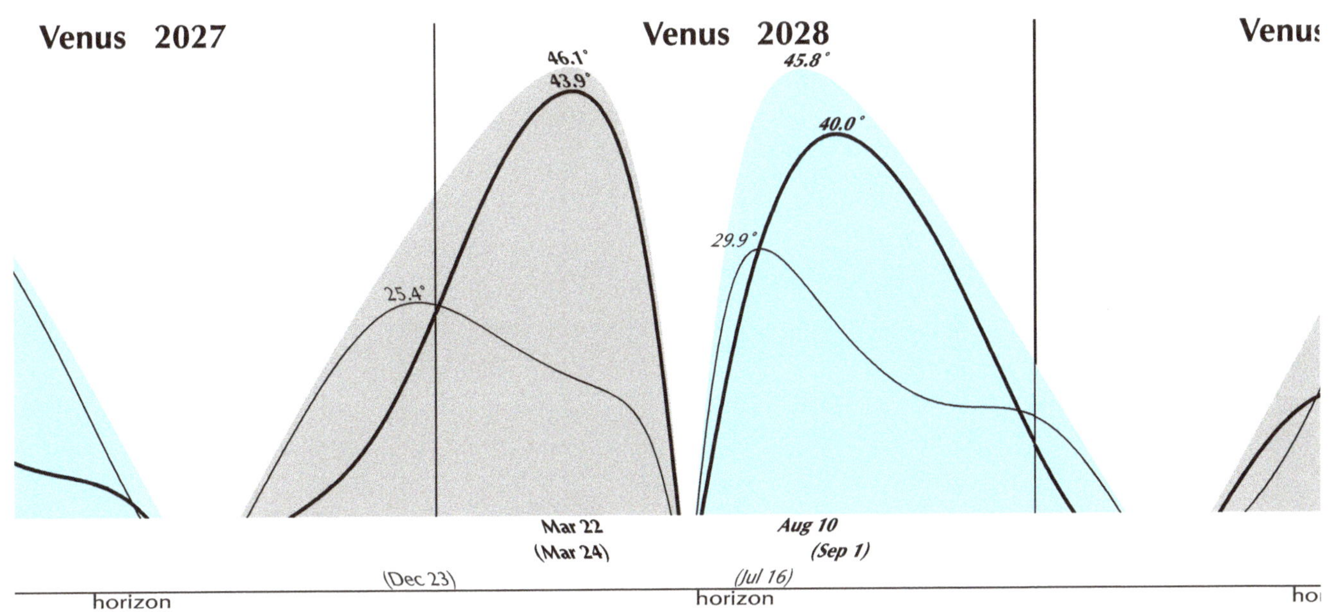
Venus 2027
Venus 2028
Venus
46.1°
43.9°
45.8°
40.0°
29.9°
25.4°
Mar 22
(Mar 24)
Aug 10
(Sep 1)
(Dec 23)
(Jul 16)
horizon
horizon

Transits of Venus over the Sun

Venus and Mercury, the two inner planets, pass between us and the Sun, but usually not exactly. But sometimes they literally get in front of the Sun's blazing disk. This is the phenomenon of the transit, which is both enthralling to observe and important in scientific history.

It is rare–much more so for Venus than for Mercury. Venus transit last happened in 2012, before that in 2004, and before that in 1882; it will next happen in 2117.

If Venus orbited in the same plane as the Earth (the ecliptic), we would see it pass across the Sun's exact center (taking about 7½ hours) at every inferior conjunction, that is, about every 584 days (its synodic period). In its real orbit, inclined at 3.4°, it transits at only about 1 in 46 of the inferior conjunctions–13 or 14 times per thousand years–and not exactly centrally.

These transits happen only around the beginning of June or the beginning of December. (Over the centuries, these dates have been shifting from May into June, and from November into December.) The reason is that, for Venus to be on a line with the Sun, it must be near one of the two nodes of its orbit: the points where it crosses our orbital plane. The line of nodes is oriented so that it points toward the early-June and early-December parts of our orbit. If, when Venus passes a node, the Earth happens to be near the same longitude of its orbit, we see a transit.

Put another way: there will be a transit if Venus comes to inferior conjunction close in time to when it comes to a node. This happened in 2004: descending node June 7, 15h Universal Time; Sun-conjunction only 18 hours later. So Venus passed across a southern chord of the Sun. Because of the eight-year periodicity, the same–approximately–had to happen in 2012.

But the eight-year cycle gradually evolves. So in 2012 the node passage had shifted later and the two events were in reversed order: Sun-conjunction June 6, 1 U.T.; descending node 16 hours later. So Venus passed across a chord of the Sun about a third of a degree farther north–north of the center and slightly nearer to it, thus taking a slightly longer time. As in 2004, Venus arrived at aphelion a little over a month later.

Though Venus's orbit is inclined to the ecliptic at only 3.4°, we see the planet cross the Sun at a steeper slope. Venus is overtaking us, but we too are moving forward, making Venus appear to move forward more slowly and therefore drop more steeply.

Since the descending node is rather near to the aphelion of the orbit, Venus during a June transit is about 0.005 a.u. farther from the Sun than at a December one. This would put it nearer to Earth; but Earth, also, is near its aphelion (about July 4), and its orbit is more eccentric. The net result is that Venus at a June transit is about 0.025 a.u. more distant from us, and appears smaller. It moves slightly more slowly across the Sun, both because it is farther from us and because near aphelion it is physically traveling more slowly.

And because it is farther from us, its angular distance from the ecliptic plane is smaller. This contributes to making the crossing time longer. And it makes Venus slightly less likely to miss the Sun. So there are more May-June than Nov.-Dec. transits (23 vs. 17, in the period 1275 to 3956, according to Jean Meeus).

Because Venus's orbit is nearly circular, these effects are much less important than they are for Mercury with its markedly eccentric orbit. But they do account for some differences between June-descending and December-ascending transits.

a pretty girl who naked is
is worth a million statues

—e.e. cummings, Complete Poems, 1913-1962

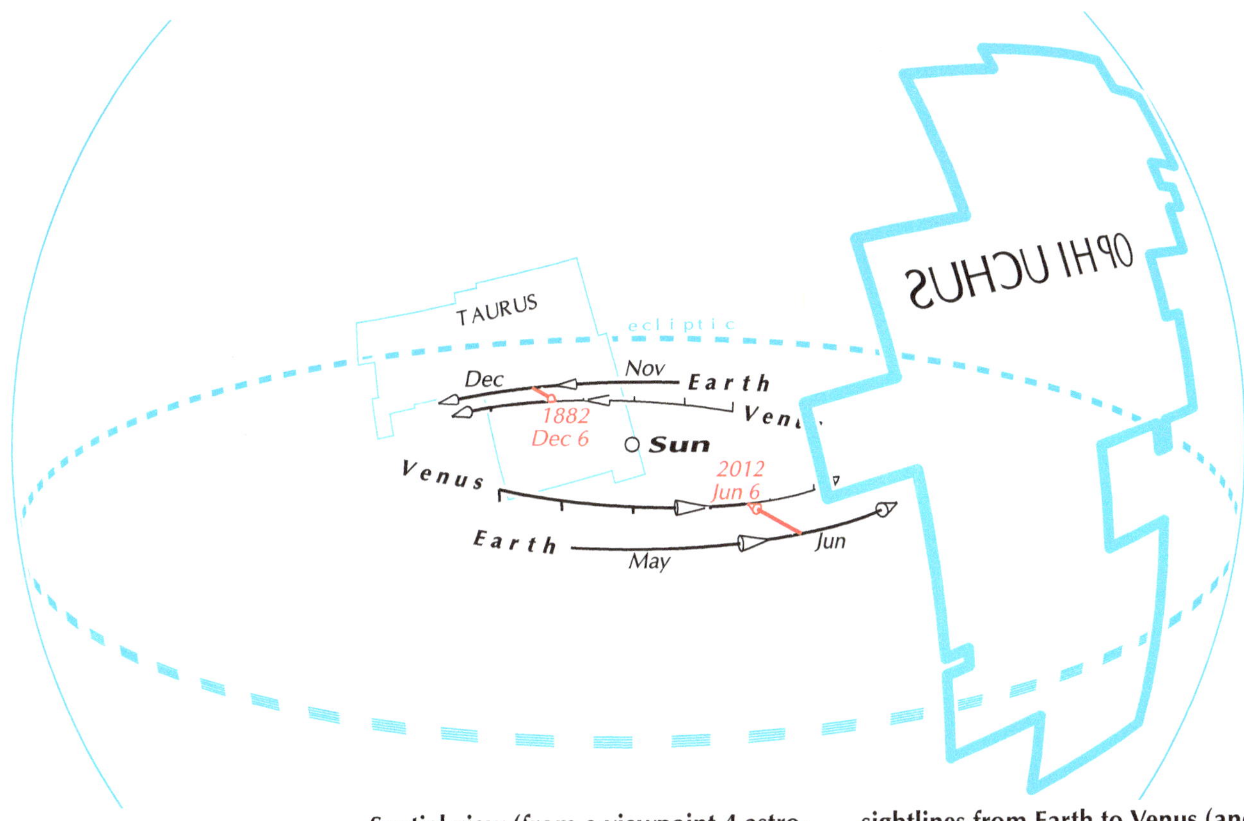

Spatial view (from a viewpoint 4 astronomical units from the Sun and 15° north of the ecliptic plane), showing the paths of Venus and Earth in November and December of 1882, and May and June of 2012. Venus's path is drawn with a thinner line after it has passed the descending node and is south of the ecliptic plane. Stalks connect it to the ecliptic plane at days 1, 11, and 21 of the months. The red arrows are the sightlines from Earth to Venus (and the Sun) at the times of the transits. They lie along the line of nodes of Venus's orbit. The boundary of constellations Taurus and Ophiuchus are drawn on the surface of an imaginary sphere 2 a.u. out from the Sun. December ascending-node transits take place against the background of Taurus, and June descending-node transits in the opposite Ophiuchus direction.

Some transits of Venus

1275 May 25
1283 May 23

1396 Nov 23

1518 May 26
1526 May 23

1631 Dec 7
1639 Dec 4

1761 Jun 6
1769 Jun 3

1874 Dec 9
1882 Dec 6

2004 Jun 8
2012 Jun 6

2117 Dec 11
2125 Dec 8

2247 Jun 11
2255 Jun 9

2360 Dec 13
2368 Dec 10

2490 Jun 12
2498 Jun 10

2603 Dec 16
2611 Dec 13

Some transits of Mercury

1993 Nov 6
1999 Nov 15
2003 May 7
2006 Nov 8
2016 May 9
2019 Nov 11
2032 Nov 13
2039 Nov 7
2049 May 7
2052 Nov 9
2062 May 10
2065 Nov 11
2078 Nov 14
2085 Nov 7
2108 May 12
2111 Nov 4
2124 Nov 15
2131 Nov 9

O Venus, queene of loves cure,
Thou lif, thou lust, though mannes hele,
Behold my cause and my querele,
And yif me som part of thi grace,
So that I may finde in this place
If thou be gracious or non.

–John Gower, *Confessio Amantis* ("Confession of a Lover," 1390). *Cure*, "remedy"; *hele*, "health"; *yif*, "give"; *querele*, "quarrel." The frustrated Lover invokes Venus; she appears and makes him confess his sins to her Chaplain; the result is a series of stories. The poem, in eight Books, was almost as popular in its time as Chaucer's *Canterbury Tales.*

The pattern of transits

The difference between the 2004 event and its 2012 successor was caused by timing: the timing of the planet's arrival at the transit-enabling part of its orbit. The evolution of the orbit causes the line of nodes to precess backward in relation to the conjunction with the Sun.

In 2004, Venus arrived at its descending node, then 18 hours later at inferior conjunction. So it transited across the Sun's southern hemisphere, taking 6 hours and 11 minutes. But in 2012 it arrived first at inferior conjunction, then 16 hours later at descending node. So it transited across the Sun's northern hemisphere, somewhat closer to the center, taking 6h40m.

But after another eight years, in 2020, Venus goes through inferior conjunction 49 hours before coming down to the descending node. So its path lies another third of a degree farther north, and misses the Sun, which is only 32' wide.

The corresponding thing happened when, 130 years earlier, inferior conjunction fell in December 1874, close enough to the other node (ascending), and there was a transit, After another eight years, the transit repeated, but 2.4 days earlier, and the node had moved forward a bit, so that the path lay about 24′ farther south. Another eight years later, the path being another 24′ farther south missed the Sun.

So these successive transits of Venus at eight-year intervals can come in groups of no more than two. Sometimes there is only one: Venus's path is close enough to the center of the Sun that the paths of both adjacent occasions–eight years earlier and later–miss the Sun.

These patches of time, within which transits (two of them or one) can occur because the line of nodes is close enough to inferior conjunction, are called by Jean Meeus "transit seasons," on the analogy of the much shorter "eclipse seasons" that happen every year for the Moon. The transit seasons fall on average 120 years apart. Rather more of them are two-transit than one-transit seasons, so altogether there are 81 transits between 2000 B.C. and 4000 A.D.: a rate of 1.35 per century.

In our age (between 1518 and 2984) all the transit seasons are double. Each group of two May-or-June transits at the descending node is followed 105½ years later by a group of two December transits at the ascending node; which is followed 121½ years later by another June descending-node pair.

But in an earlier age (427 B.C. to 424 A.D.) there was a succession of single-transit seasons (alternately May-descending and November-ascending), with gaps of 121½ years between them. Other ages are transitional, with mixtures of one-transit and two-transit seasons.

The diagram shows many of the pairs, such as the transits of 1275 and 1283, 2004 and 2012 . . . Or at the ascending node, 2603 and 2611. Among the transits shown, 1396 Nov. 23 is the only one passing close enough to the Sun's center that the corresponding events 8 years before and after were not transits: at the inferior conjunctions of 1388 Nov. 26 and 1404 Nov. 21 Venus passed respectively just too far north and much too far south. So this was a one-transit season, the nearest to our time–there will not be another till 3089 Dec. 18.

The diagram also shows that the transit seasons at one node, whether giving rise to two transits or one, recur almost exactly 243 years later: the 1275-1283 pair is repeated in the 1518-1526 pair, only about 1′ farther south. This is because 395 Venus-circuits around the Sun equal 243 Earth-circuits, even more closely than the 13-8 harmony. Still not precisely, or the transit paths would repeat precisely; instead they work their way southward across the Sun's face.

They thus form long series (reminiscent of the famous Saros series of similar eclipses): 1283 repeats almost identically in 1526, then in 1769 . . . Four such series are in progress in our age, and are designated B, C, E, and F. B and E are running the end of their long courses; C and F are chasing them across the Sun, having started relatively recently. But at some other epochs there have been only three or two series current. Each series consists of up to 20 members and thus lasts up to 4860 years.

Actually 2611 is not quite the last of its series: on 2854 Dec. 14, though Venus misses the Sun as seen from the center of the Earth, it will briefly touch as seen from part of Antarctica. Grazing transits like this occur only at the beginnings and ends of series; in some the transit is complete (planet appearing entirely inside the Sun's disk) from only a fraction of the Earth's sunward face, and partial (merely indenting the Sun's edge) from another fraction. (Such a partial transit happened for Mercury on 1999 Nov. 15.)

Transits of Venus, 1275-2611. The planet is shown to scale (not exaggerated) , at intervals of an hour. The figure after each date is the distance in degrees between the centers of planet and Sun, at mid transit. Red emphasizes the two transits of the 21st century, and the next two passages of the same cycle that miss the Sun.

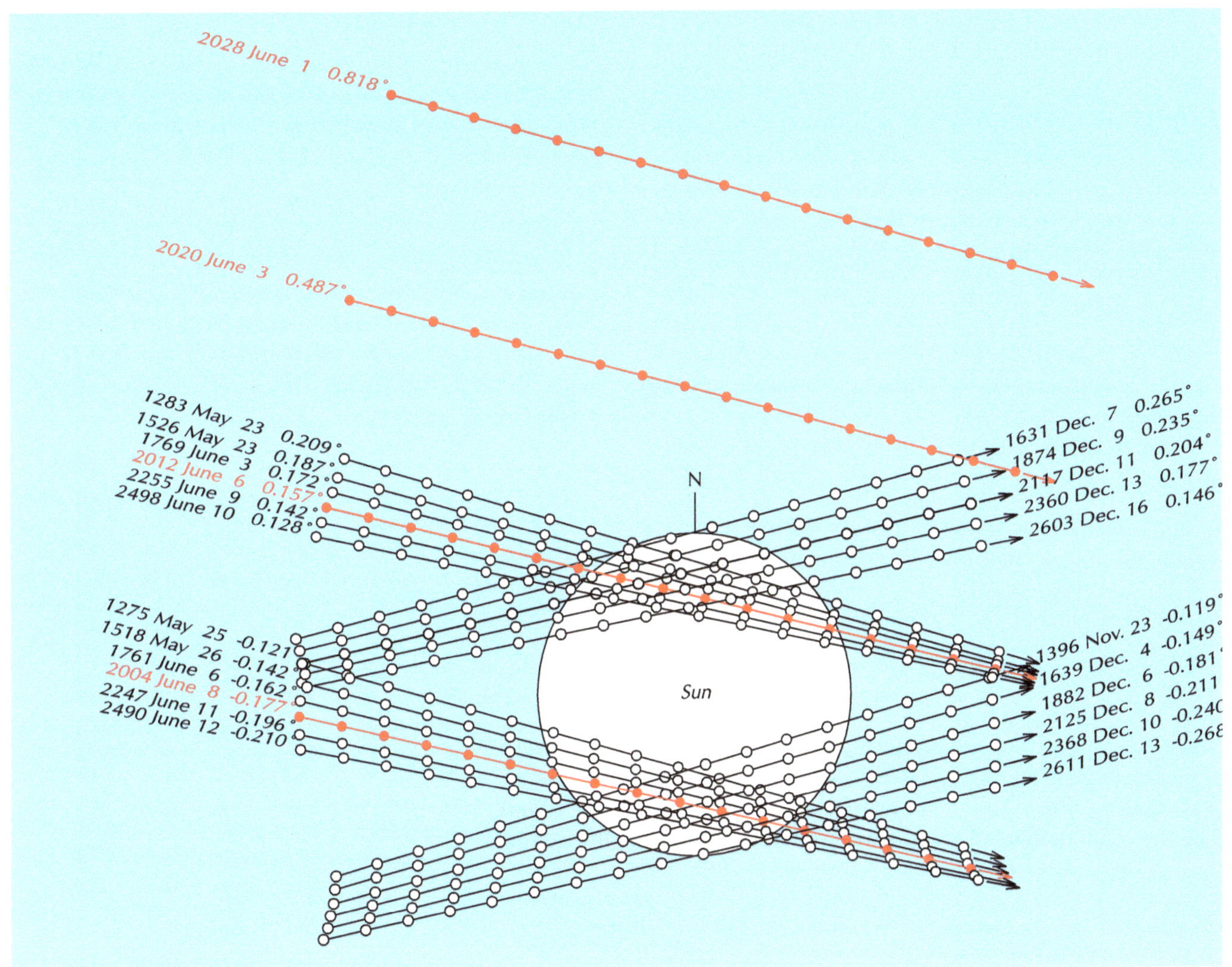

Rise and hold up the curved glass
And pour us wine of the morning, of El Andar . . .
By herself she is fearless
And gives her arms to the air,
The limbs of a long camel that has not borne.
She gives the air her breasts,
Unfingered ivory . . .

–Opening of the *mu'allaqah* of 'Amr ibn Kulthûm, freely translated by Powys Mathers in *Love Songs of Asia*, a collection edited by his widow in 1944. I thought I remembered that this little book had a section "Songs from the Love Nights of Laos," but that must have been somewhere else. The seven *mu'allaqât* were long poems "hung up" in a temple after winning the prize at an annual fair, and form a Pléiade of pre-Islamic Arabic poetry.

The most recent transits

2004 and 2012 were the only Venus transits of our lifetime.

In 2004 I and friends saw the transit from a hillside called Heaven's Gate, overlooking the Lym valley, on the border of Dorset and Devon. In 2012 we were on the wrong side of the Earth. Many enthusiasts traveled great distances to observe these spectacles through telescopes.

The area of Earth from which a transit is visible is determined mostly by the time of day, also by the season of the year. In June, our north pole is tipped toward the Sun and so the region of observability includes the Arctic and does not reach to the Antarctic. In November, approximately the opposite.

The hemisphere of observability is centered on the longitude at which the Sun is overhead at mid transit.

On 2004 June 8, that spot was in eastern Arabia (just about at the meeting of Saudi Arabia, Oman, and the United Arab Emirates). So the middle of the event favored all of the Old World–Europe, Africa, Asia–and (low in the sky) Greenland and part of Australia. Earlier stages were seen in eastern parts of the Americas, later stages in more of Australasia.

In 2012 the central sub-solar spot was in the west central Pacific. So the hemisphere then facing Venus and Sun included most of the Pacific, Australasia, part of North America, and a larger part of Asia. Because the International Date Line runs through the Pacific, the event began on June 5 but centered and ended on June 6.

The hemispheres from which the beginning, middle, and end of the transits can be seen. An arrow shows the distance Earth advances in 1 minute; an arrow on the equator shows how far it rotates in 1 hour. The International Date Line follows, mostly, the 180° line of longitude.

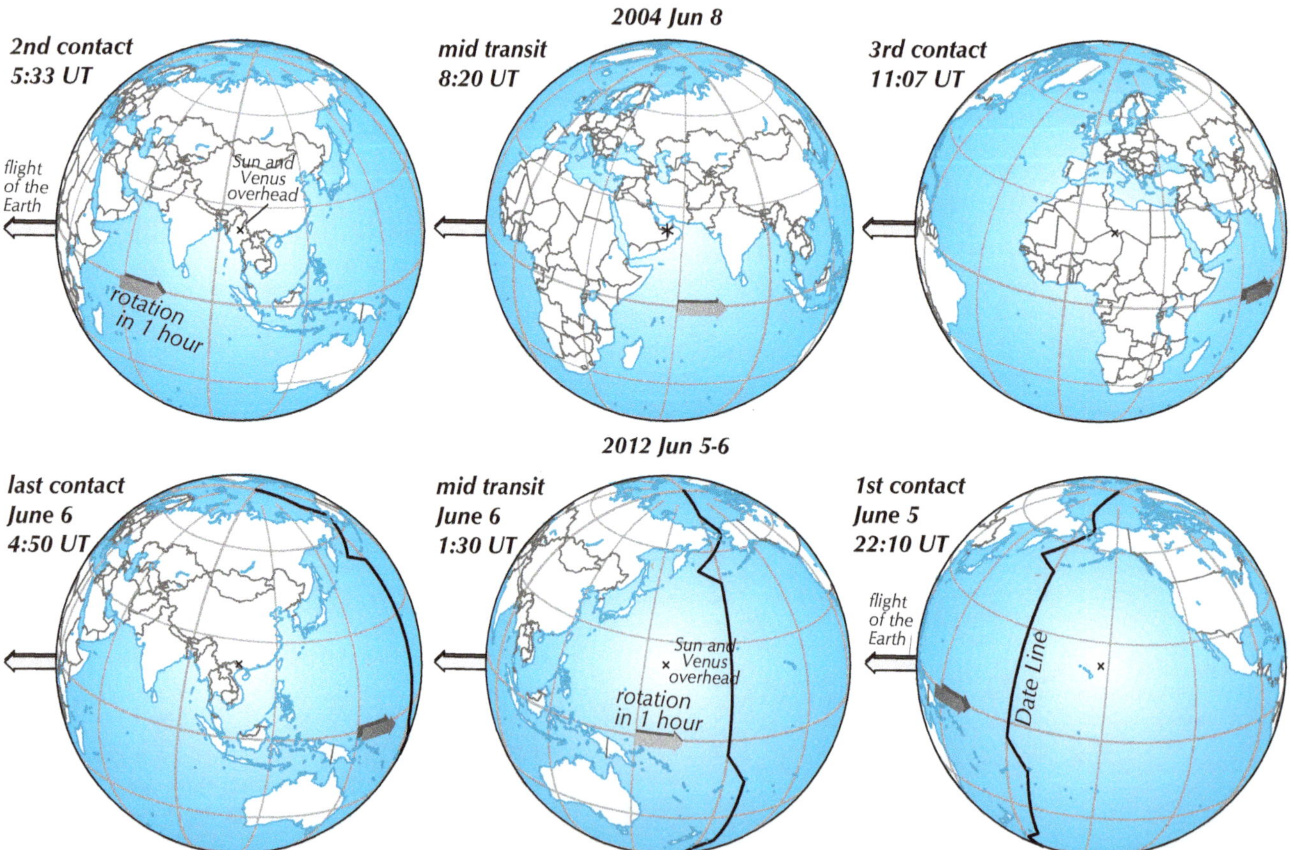

The phenomena of 2012 June 5-6

Venus was 0.289 astronomical units from Earth, and 0.726 a.u. from the Sun. The Sun, 1,392,000 kilometers in diameter, appeared 1,891 seconds wide (about 0.53°). Venus, 12,100 km in diameter, appeared 58 seconds wide–nearly 33 times smaller. Yet it was wide enough to be discernible by the naked eye, if any naked eye had been foolish enough to try to look at it! At mid transit this black spot was 391 seconds in from the Sun's edge, 554 from its center.

Here are the Universal Times of mid transit and of the four instants called "contacts," when the apparent disk of the planet was tangent to the limb (outline) of the Sun, on the way in (ingress) and out (egress).

	h m s		
June 5	22:09:41	1	ingress, exterior contact
	22:27:29	2	ingress, interior contact
June 6	01:29:36		mid
	04:31:43	3	egress, interior contact
	04:49:31	4	egress, exterior contact

These are the geocentric times (as seen from the center of the Earth. For actual localities on the surface the times can be up to 5 minutes different–later if west of the midline of Venus's "shadow" on the Earth. This is a parallax effect: if you are farther west, Venus appears slightly farther east, and seems to hit the Sun later.

And since Venus, this time, crossed north of the Sun's center, if your viewpoint was northerly you saw it cross a slightly longer chord of the Sun's disk. This is the parallax effect by which earlier astronomers hoped to find the real distance of Venus, and of the Sun.

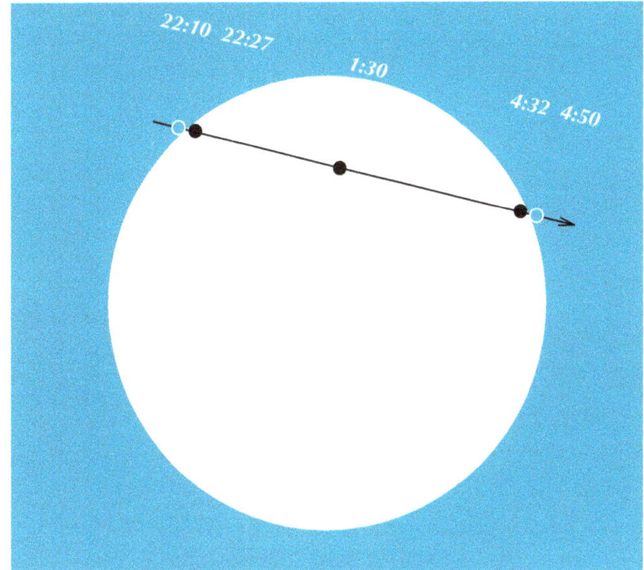

1st and 2nd contact, mid transit, 3rd and 4th contact.

Local circumstances in 2012

No transit was seen from regions turned away into night: almost all of Antarctica (in its midwinter 24-hour night); most of South America; most of the Atlantic; southwestern and western Africa; Portugal and southern and western Spain.

The whole course of the transit was seen from places that stayed on the sunward side of the Earth: the Arctic, Alaska, northwestern Canada, Greenland except the southern tip, Iceland, northern Scandinavia, northern Russia, Siberia, Mongolia, much of China, Japan, the Philippines, northern and central and eastern Australia, New Zealand except for slivers of the east shore of both islands, and the western and northwestern Pacific including Hawaii.

To the west, the transit was already in progress as the Sun rose: a zone stretching from Britain through most of Europe, all of the Middle East, northeastern and eastern Africa, central Asia, Tibet, India, southeast Asia, most of Indonesia, western Australia, just about all of the Indian Ocean, and a sliver of the coast of Antarctica. So only the stages after the beginning were seen. The farther east you were in this zone, the more you saw: so Russia and India and western Australia were up when the transit had just started, Britain and France and Madagascar came around into view when it had almost ended.

In another diagonal zone, on the east, from southern Greenland through most of Canada (eastern and central and southern), all of the contiguous U.S. and Central America, northwestern South America, and the eastern and southern Pacific, the transit began toward the end of the day, and during it the Sun sets. For points along the eastern edge of this zone, such as Bermuda and Venezuela and northern Peru, the Sun set as the transit was starting; for the western edge, such as British Columbia and the eastern shore of New Zealand, all but the end of the transit was seen.

Observing the "Black Drop"

Observers in 2004 and 2012 were seeing something that had last been seen in 1882. There are 13 to 14 transits of Mercury per century, and to see one of those is awesome enough. What photographs show as a mere black dot, the mind feels as a far-off ponderous sphere sailing before a far vaster one. If that is so for Mercury, it is more so for Venus, much nearer and larger.

We'll describe what your experience was if you observed in 2012.

Besides absorbing the strange spectacle, you may have been one of those who timed it (using a stopwatch corrected by a radio time signal), and thus perhaps added to the long effort to refine scientific knowledge.

The moment of first contact was difficult to judge. The planet, as yet an invisible part of the sky, had to dent the Sun's limb, and this could hardly be noticed until some seconds after it had happened. Then you had to wait through 18 minutes during which the 58-second-wide black spot slowly entered the blazing 1891-second-wide disk.

Venus, unlike Mercury, has a thick atmosphere, tending to make timings fuzzier. Sunlight shines through, so that when the disk is still part way onto the Sun a luminous arc is discoverable just outside the Sun's limb.

There came a stage when you thought the planet was all inside, tangent to the edge. But you had to wait some more long, suspenseful seconds. The black spot became slightly stretched toward the limb; then seemed to hang by a short thick stalk from the great black curve, like a "black drop."

This deformation is apparently caused by irradiation (spreading of brighter light in the eye) but may also be affected by turbulence in our atmosphere, since it is more marked when the limb of the Sun "boils." Suddenly, the thread breaks: Venus is completely if infinitesimally surrounded by light. The instant of the breaking was the one to time as second contact.

Similarly, just over 6 hours later, at third contact: the planet squeezed to a slightly oval shape, reached out to touch the Sun's limb, suddenly became stuck to it by a thread–you pressed your stopwatch at that instant: third contact. Then for nearly 18 minutes it swam out, and judging exactly when the last dent disappears was difficult. There could be a spread of as much as twenty seconds among people's timings.

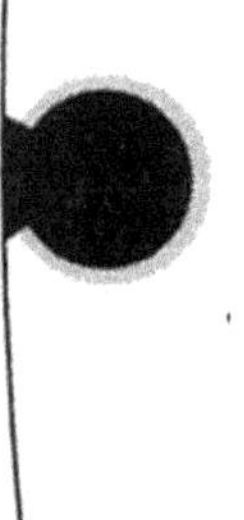

Captain Cook's drawing of one of the black drop stages in the transit of Venus, 1769

A 2004 photograph

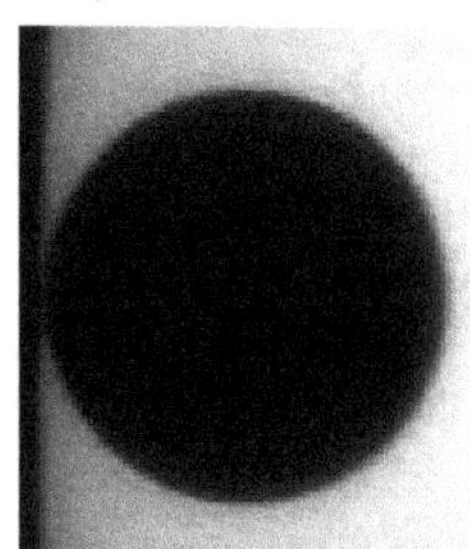

Looking at the Sun is dangerous!

For the 2004 and 2012 transits, we had to give this stern warning. Bear it in mind if you're alive in 2117 and in the even more unlikely case that technologies have not changed.

TO AVOID BEING BLINDED, DO NOT LOOK AT THE SUN with naked eye or with any false protection such as sunglasses or smoked glass or colored water or exposed film! Make a pinhole in a piece of card, let the sunlight fall through it onto a large white card, and adjust till you see a sharp inverted image. Or look through a piece of welder's glass, strength 14 (or 12 if the Sun is low and dim).

Look through a telescope ONLY if you know exactly what you are doing. The safe solar filter must be at the front (entirely covering the entrance aperture)–not at the back (behind the eyepiece) where focusing the Sun onto the eyepiece could suddenly crack it. You can also use a telescope or binocular to project the Sun's image onto card.

History—Kepler, Gassendi

2012 was only the 8th transit of Venus since the invention of the telescope in 1609. It is just possible that earlier ones were noticed by Japanese observers, who used to look briefly at the very low Sun.

Johannes Kepler estimated wrongly that Mercury would transit the Sun on 1607 May 27, and thought he observed it, with light projected through a hole in a wall onto paper. Ten years later, after Galileo had started the application of telescopes to astronomy, and, among many other discoveries, had seen sunspots, Kepler realized that this was what he had been the first to see.

He did correctly predict that, amazingly, in 1631 both Mercury and Venus would transit, on November 7 and December 7! Alas, he died in 1630.

The Mercury transit was observed by Pierre Gassendi from Paris. It happened in the morning, some hours earlier than predicted. And Mercury was 6 times smaller than was then expected, so that Gassensi might have missed it if he had not been projecting the Sun through a telescope. And the surprisingly small size made his contemporaries skeptical. But this was the first transit, of any planet, definitely observed, and it proved the general accuracy of Kepler's system–his famous three laws of planetary motion.

Gassendi tried for Venus a month later, but failed because the prediction was off. We now know that the transit ended half an hour before the Sun rose at Paris. This was a sore disappointment, because the next Venus transit predicted by Kepler lay more than a century ahead.

Horrocks

Jeremiah Horrocks is the most poignant of the young geniuses in this story. Born about 1618 at Toxteth Park near Liverpool, he supported himself by tutoring, and in 1639 became curate of Much Hoole, about six miles southwest of Preston in north Lancashire. Having taught himself astronomy, he studied Kepler's tables, correcting them from his own observations "with a half-crown telescope" (Henry C. King, *History of the Telescope*; the half crown was a coin worth two shillings and sixpence, or an eighth of a pound). He calculated that what would happen on 1639 Nov. 24 (in the Julian "Old Style" calendar; Dec. 4 Gregorian) would be not a near miss but another transit.

The day was a Sunday. Horrocks had alerted a friend, William Crabtree, at Broughton near Manchester, but he himself, between church services, had to rush to the darkened room where he had made careful preparation. He had the Sun projected onto a huge circle which he had painstakingly marked with measures. and which he must have had to move as the Sun moved. Thus he became the first to witness any Venus transit–along with Crabtree, to whom weather allowed only one glimpse.

From the size of Venus's projected disk, Horrocks figured that, as seen from the Sun, it would appear 28 seconds wide, and that this gave the clue to the sizes of the planets and their orbits. The logic was wrong, but the result was better than any previous: he got the Sun at 63%of its real distance.

The other astronomical work of his short life, especially on the orbit of the Moon, anticipated Isaac Newton. He died at Toxteth on 1641 Jan. 3, aged about 22. Many of his papers were destroyed by carelessness and the English civil war that broke out in 1642.

But his precise fixing of a few points in the 1639 transit fixed the predictions for all future transits. A copy of his manuscript Venus in Sole Visa ("Venus Seen in the Sun") reached the eminent astronomers Christiaan Huygens (in Holland) and Johannes Hevelius (in Danzig) and inspired them to observe the next transit visible from Europe, that of Mercury in 1661.

Gregory

James Gregory, a 25-year-old Scot, suggested in his Optica Promota of 1663 that transit timings might be used to settle what had become astronomy's huge unanswered question: the Earth-Sun distance. This is a quantity so fundamental for the planetary system that it is called the "astronomical unit."

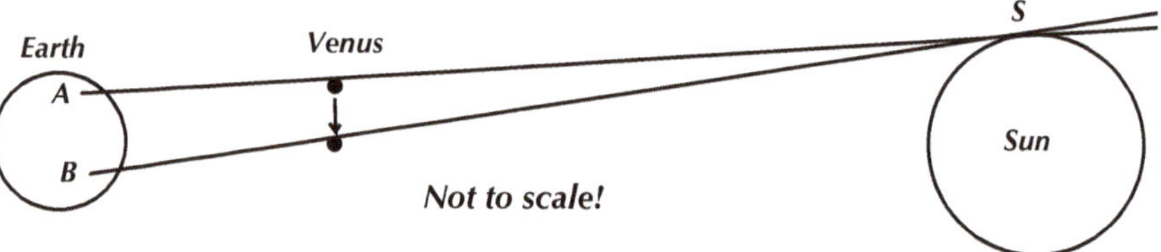

The relative positions of the Sun and planets were now known, thanks to masses of angular observations of their positions, and the explanation of these by Kepler's laws. But not the absolute distances among them. Those would all fall into place once the astronomical unit could be determined.

The situation was rather like looking out of a window and not knowing whether those figures on the street are giants or dwarves because you don't know the length of your garden path.

We could triangulate the Sun's distance if, from places on opposite sides of the Earth, we could take sufficiently exact sightlines to it against the background of the stars. This angle is the "solar parallax"; 1 / sin parallax = the Sun's distance in units of the radius of the Earth. Doing this directly is impracticable because the Sun hides the stars.

But if observers at A and B time the different instants when, for them, Venus begins to transit the Sun, then the angle ASB between their sightlines can be found from the angular distance Venus is known to travel in that time; thus, from distance AB and angle ASB, the lengths of the triangle's sides can be found.

This method depends on having clocks well synchronized, in locations far apart. The method later suggested by Edmund Halley uses the different durations of the transit as seen by observers at different latitudes. The difference between the durations, as seen from places a known north-south distance apart on the Earth, gives the angular distance between the chords across the Sun. It will be smaller as the distance from Earth to Venus is greater.

Gregory also in 1663 invented the first reflecting telescope, the "Gregorian" type that was to be used by Cook. In it the secondary mirror sends the image back through a hole in the primary to an eyepiece lens at the rear. The now more familiar "Newtonian," with eyepiece to the side, was first made by Newton in 1668.

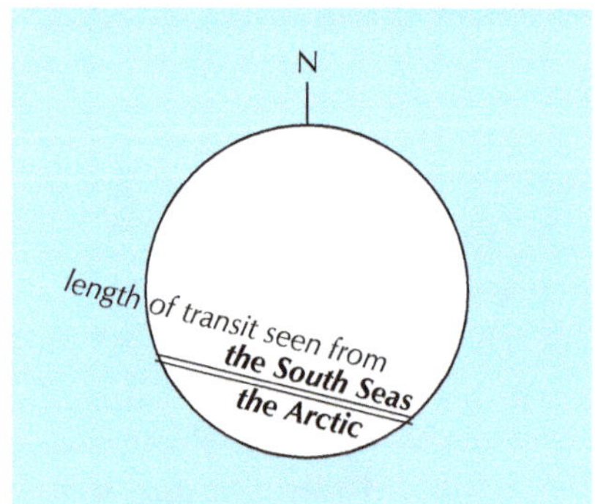

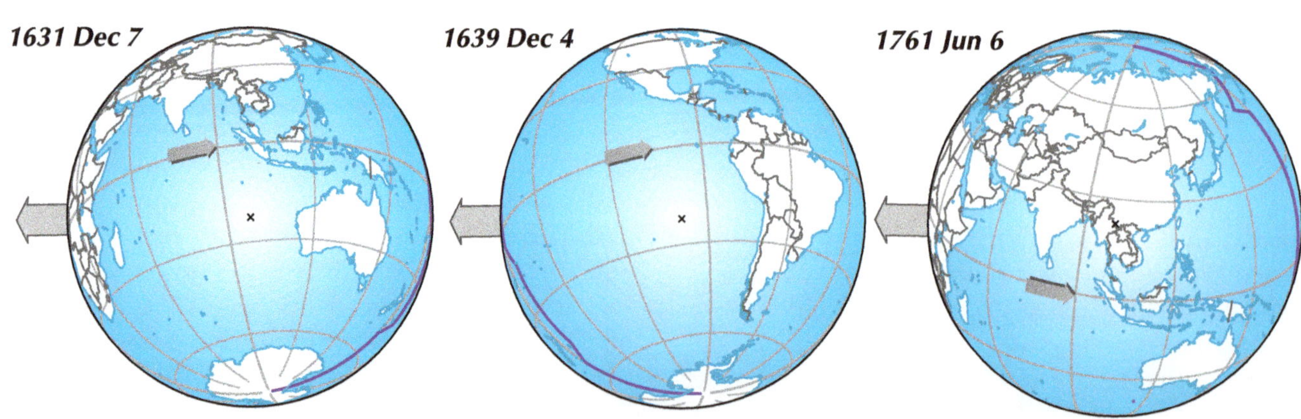

Halley

Britain under King Charles II, in competition with France, made an investment in astronomy so as to determine position for ships at sea. Central to this was the founding of the Greenwich Observatory in 1675, and the choice of it for the prime meridian of longitude.

In 1676, Edmund Halley at the age of 20 got himself sent by the king to catalogue the southern stars from the island of St. Helena in the South Atlantic. At the Mercury transit of 1677 Nov. 7, he managed to see the contacts despite clouds. He described the "black drop" effect, and realized that it yielded timings of one-second accuracy, which he was the first to make.

He may or may not have read Gregory, but saw the application to the distance of the Sun, and hoped that someone in Europe had been able to make timings. No one had, except for a poor one by Jean-Charles Gallet, a churchman at Avignon (who however noticed the deformation of the planet's disk toward egress).

Halley later realized that the method would have been too difficult with Mercury. So he urged the scientific community to prepare for that which would happen long after his own death and would be "by far the noblest [sight] astronomy affords": a transit of Venus.

Accordingly there was intense interest in the far-off next pair of Venus transits. In 1761 France and Britain were at war, which was part of the reason why many of the expeditions they sent to remote places failed. Those that succeeded yielded Sun-distances between 123,000,000 and 157,000,000 kilometers.

1769 would really be the last chance for a century, and every effort was made to observe it. It was for this purpose that George III had the King's Observatory built, between Richmond and Kew, west of London. The longest duration of transit would be seen from Scandinavia and the shortest from somewhere in the South Seas. Expeditions were sent to California, Santo Domingo, Hudson's Bay, northern Norway, Lapland, and India. The most extraordinary was the three-year voyage of Lieutenant (later Captain) James Cook.

Venus transits over five centuries. Each picture shows the Earth hemisphere in view of the middle moment of the transit. A large zone farther east sees the earlier stages, and a zone to the west sees the later stages.

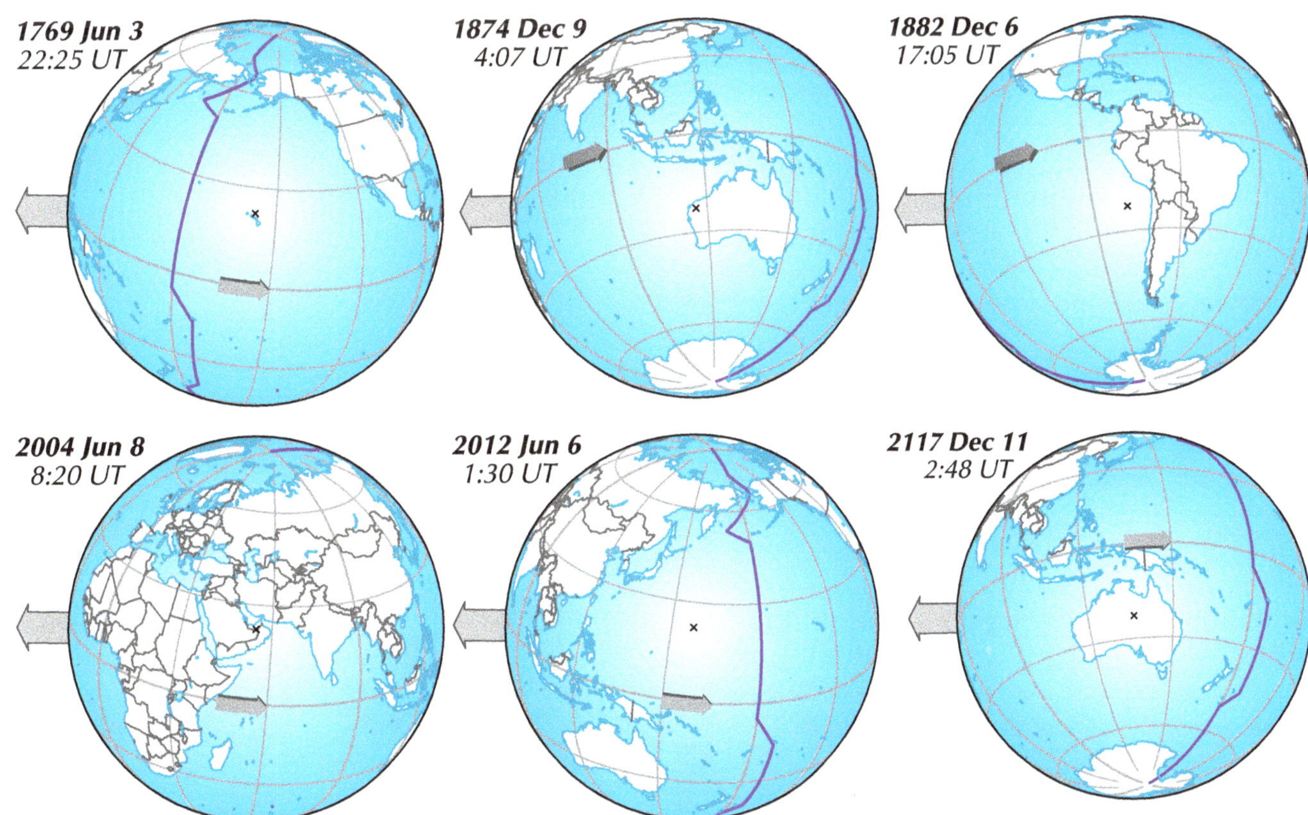

Cook's voyage

Born in 1728 to a Yorkshire farm laboring family, he rose through navy ranks, becoming skilled in mathematics and surveying on voyages to the Baltic and America. He made such accurate observations of a solar eclipse in 1766 that he was chosen two years later to command the ship "Endeavour" on a mission for the Royal Society and King George III—who himself saw the transit from an observatory built for him at Kew.

Cook and astronomer Charles Green were to catch the transit at Tahiti. He was also to extend the geographical discoveries of Abel Tasman, Samuel Wallis, and John Byron. With him were Joseph Banks and Daniel Solander, to record natural history and ethnology.

Cook sailed from Plymouth in August 1768, reached Tahiti in April 1769, and on June 3, at Point Venus on the island's northern end, Cook, Green, and Solander did their best to make independent timings of the transit. The telescope probably used by Cook, a Gregorian reflector made by James Short, is on display in the New Zealand museum at Wellington. They were unprepared for a problem not encountered at the transits of Mercury: "We very distinctly saw an Atmosphere or Dusky shade round the body of the planet," so that "we differ'd from one another in observing the times of the Contact much more than could be expected"—by as much as 20 seconds.

Tasman had in 1642 discovered land that he called Nieuw-Zeelamd, after the Dutch province of Zeeland—a brief landing on the northwest coast of the South Island and a fight with natives—and this, it was thought, might be the west coast of a huge southern continent. So Cook sailed south to an icy latitude 40°, without finding the supposed continent, then turned west to New Zealand, and circumnavigated both islands, charting them with astonishing accuracy.

It happened that 1769, like 1631, was a rare year when both Mercury and Venus transited. So on the morning of November 9 Cook, Green, and Lieutenant Zachary Hicks observed the Mercury transit from the Coromandel Peninsula, on the eastern side of North Island, at Mercury Bay, as Cook called it. Its native name, according to the log book kept by Hicks, was Oporege, "oyster river," though this is impossible in Maori. The log book pages for the observation are lost, and Green and Hicks died before the end of the voyage. It has been argued that Cook, being now in the southern hemisphere, made the elementary mistake of starting to look for Mercury on the wrong side of the Sun, failed to time the transit, and covered it up by "losing" the record and pretending that he himself was not timing the transit but measuring the position of the Sun. This seems far out of character.

(For more on Cook at Mercury Bay you can see the cover painting and story for my Astronomical Calendar 2004.)

Cook went on to chart the west coast of Australia, bring the "Endeavour" safely home via the Cape of Good Hope. He made two more enormous voyages of exploration all over the Antarctic and Pacific Oceans and through Bering Strait into the Arctic.

Besides such achievements as demonstrating the use of clocks in fixing longitude (for which he was academically rewarded), he was an unusually approachable and respected Captain, in an age when conditions for sailors were generally brutal. He showed the value of keeping his sailors healthy with fruit and vegetables, formed friendships with chieftains, and did his best to obey the enlightened "Hints" of the Royal Society's president, James Douglas Earl of Morton: "To exercise the utmost patience and forbearance with respect to the Natives . . . They are human creatures . . . the legal possessors of the Regions they inhabit . . ."

But endless trouble with natives who thieved, and sailors who stole the ship's nails to pay Polynesian women for sex, led Cook into more and more floggings of sailors and shootings of natives. In 1779 he was killed, in a riot on the beach of one of his discoveries, Hawaii. His wife, who had not seen much of him, died fifty years after.

Cook's telescope

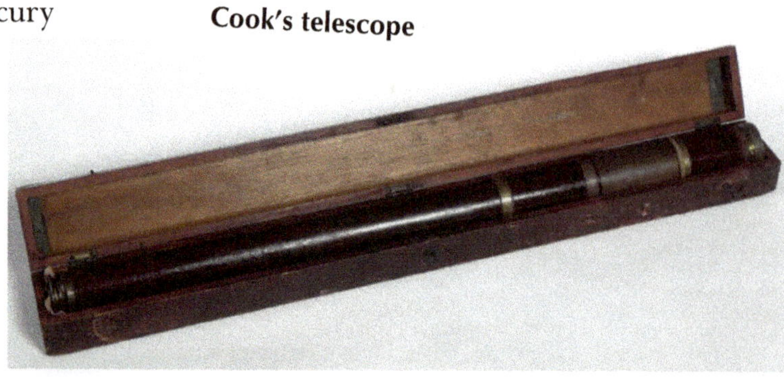

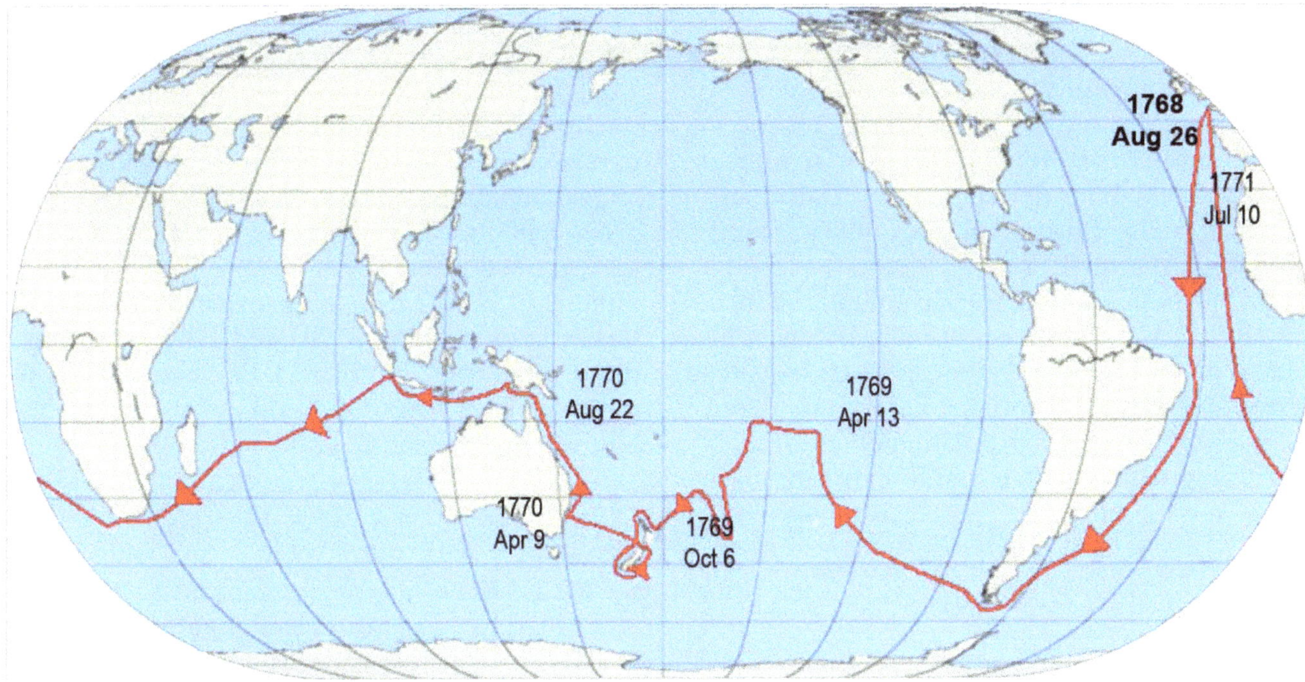

Cook's 1768-1771 voyage.

Young James Cook was a sailor bold,
He was brave, he was good, he was clever;
He rose to be captain of the king's navee
And commanded the good ship "Endeavour."
He said to his wife "You're the joy of my life
Though oceans roll between us,
But I must be off to the isles of the south
To observe the transit of Venus."

He sailed away out of Plymouth Bay
With his doughty crew beside him,
With a ready sail for breeze or gale
And the faithful stars to guide him.
They sailed that way for many a day,
That gallant band of freemen,
Far to the south in the tempest's mouth
Till they passed Cape Maria Van Diemen.

Then "Land Ahoy!" cried the cabin boy
And happy they were to see land,
And going ashore raised the flag they bore
On the bush-clad hills of New Zealand.
As they made their claim in their sov'reign's name
They heard the bell-bird singing,
And the news they did send at their journey's end
Set the bells of old England a-ringing.

Willow Macky (1921-2006), New Zealand folk-singer. She was commissioned to compose it for the 200th anniversary of Cook's 1768-1771 voyage. I know the song only from hearing a New Zealander sing it. My representation of the tune is with a keyless notation I use: tones 1, 3, 5, and then the upper 1.

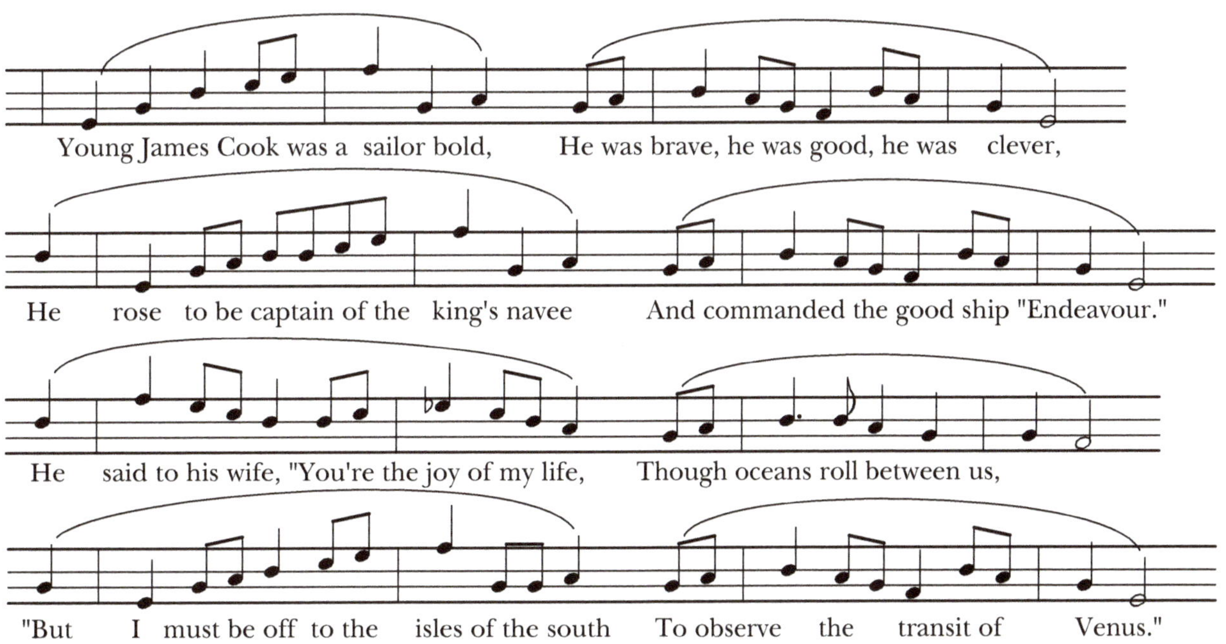

After Cook

Cook's and other transit timings differed so widely that the resulting values for the distance of the Sun, though the best yet, still ranged between wide bounds. It was too difficult to make timings accurate enough for determining such small angles. By the 1874 transit, photography was available, but it did not help.

So the hope of using transits to find the solar parallax and the Earth-Sun distance is now of historical interest only, and other methods have been found: aberration of starlight, radar sounding to Venus, gravitational perturbation of asteroids, and triangulation to Mars and to Earth-approaching asteroids such as Eros, (which approached Earth in 2012. In 2012, for only the fourth time in the history of the post-Renaissance world, we were in one of those "seasons" for which astronomers longed and prepared so ardently, when Venus gives us not one but two chances to watch her dare to tread across the Sun like a fire-walker. The number of people managing to see this in 1631 was zero; in 1639, two; 1761 and 1769, on the order of a hundred; 1874 and 1882, thousands; and probably, if we count television as "seeing," it was millions in 2004 and 2012.

Someone who saw the 2012 transit could by surviving for 105 years see the 2117 one. Joseph Shuster told me (in a comment, 2019 Oct. 14, to a blog post of mine) that this was why he coached two five-year-olds, his nephew and the son of a friend, to see the 2012 transit in a telescope, hoping that they might be the first humans to witness Venus transits of different pairs. His reasoning was that those younger than five would probably not remember the first experience, those older would have less chance of surviving to the second.

The Sun occults Venus

At superior conjunctions, Venus passes behind the Sun. At each 5th of them occurs the unobservable phenomenon that is the reverse of transit: the Sun gets in front of Venus, and, being so much larger, occults it.

This also has to happen once in each eight-year cycle, half way between the transit-producing inferior conjunctions, hence also in June: 2000 June 11, 2008 June 9, 2016 June 6, 2024 June 4, and so on.

But, unlike the transits on the other side of the orbit, the occultations happen every time, because at superior conjunction the planet is 6 times as far away, so its latitude-difference from the Sun appears small.

A book of Verses underneath the Bough,
A Jug of Wine, a Loaf of Bread--and Thou
Beside me singing in the Wilderness--
O, Wilderness were Paradise enow!

–Edward Fitzgerald, *Rubáiyát of Omar Khayyám*, translation of the *rubâ`iyât* (quatrains) of Omar Khayyân. This was Fitzgerald's 1889 version of this stanza; it differs from his 1859 version, and has a rather loose relation with two of the Persian stanzas. Omar Khayyam (1048-1131) was an important mathematician and astronomer; he devised a calendar, called Jalali, with a 33-year intercalation that makes it more accurate than our Gregorian. The poetry attributed to him was made famous when Fitzgerald's work appealed to orientalizing fashion.

Put another nickel in,
In the nickelodeon,
All I want is having you
And music, music, music!

–Stephen Weiss and Bernie Baum, 1949

Transits of Mercury

Transits of Mercury are 10 times more frequent than those of Venus: 13 or 14 a century.

Mercury is almost twice as near to the Sun, and we see it pass the Sun nearly 7 times more often. Slightly offsetting that, its orbit is more steeply inclined, so at inferior conjunction it spends less time close to the ecliptic.

The transits fall in May or November, at Mercury's descending and ascending nodes. The last was 2019 Nov. 11. The next pair will be 2032 Nov. 13 and 2039 Nov. 7.

Of the six historic modern years containing Venus transits, two, as we have seen, also contained Mercury transits. But this, despite the abundance of Mercury transits, is rare. The next double-transit years are far ahead. According to Jean Meeus, Transits (1989), p. 24:

```
1631 Nov  7 Mercury
1631 Dec  7 Venus

1769 Jun  3 Venus
1769 Nov  9 Mercury

2611 Nov 16 Mercury
2611 Dec 13 Venus

2854 Nov 23 Mercury
2854 Dec 14 Venus

2976 May 21 Mercury
2976 Jun 16 Venus
```

It's almost as if Nature arranged the exceptional order and spacing in 1769 so as to give Cook his chance to observe at Point Venus in Tahiti and then Mercury Bay in New Zealand.

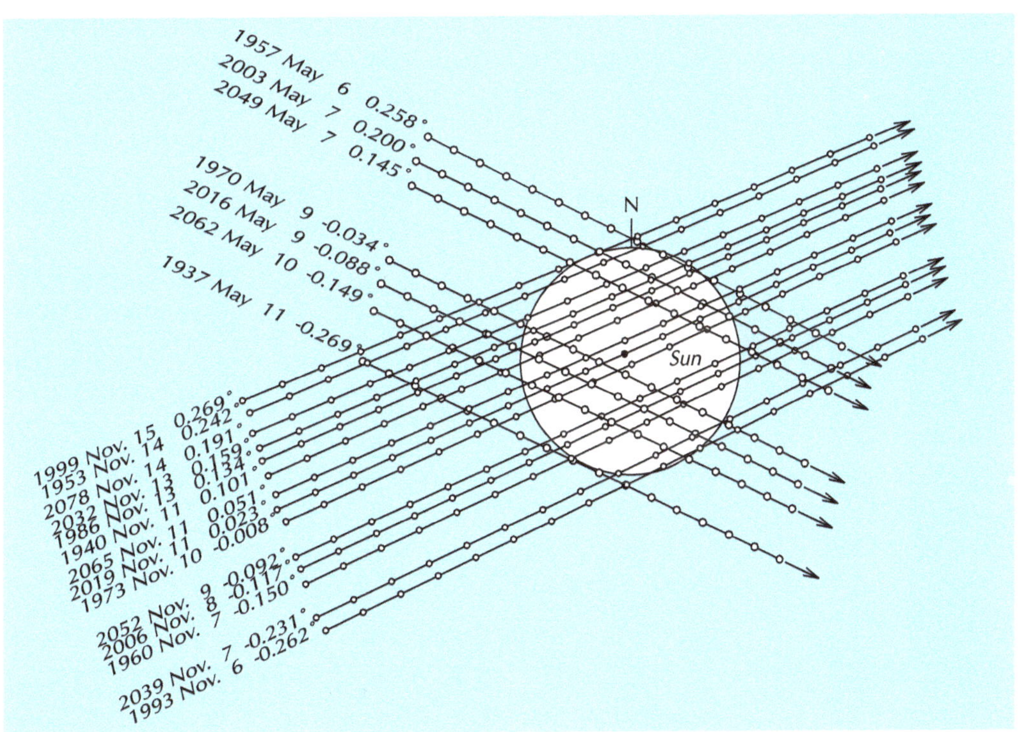

Transits of Mercury 1937-2078. The planet's size is exaggerated 5 times (whereas in our corresponding diagram for Venus there was no exaggeration of size).

More information about transits

Among other mathematical curiosities set out in Jean Meeus's Transits are listings of transits as seen from other planets. Surprisingly, transits of Venus as seen from Mars are commoner than those seen from Earth, because of the orientations and periods of their orbits. Meeus has more about transits in his *Mathematical Astronomy Morsels III* (2004), p. 264-304.

Other books: Eli Maor's *June 8, 2004: Venus in Transit* (2000); David Sellers, *The Transit of Venus* (2001).

Norman Davidson's *Astronomy and the Imagination* has interesting figures and sidelights on transits. He tells us that Alpetragius wondered why he had never seen Mercury cross the Sun's face, and deduced that the planet must shine by its own light.

Conjunctions

A conjunction ("joining") of two heavenly bodies is the event of their passing each other. They may be moving in opposite directions, or one may overtake the other, or one may be a stationary object such as a star. But there are several kinds of conjunction.

Conjunction in right ascension happens when one body is exactly north of the other in the equatorial frame of reference. Conjunction in longitude is when one is north of the other in the ecliptic system. The standard meaning now is the equatorial one. Earlier centuries, and astrology, talked in ecliptic terms. Because planets move generally parallel to the ecliptic, the angular separation at the conjunction in longitude is generally closer than at that in right ascension.

Thirdly, there is the appulse, or closest moment, when the angular separation is at minimum. The three events—conjunction in right ascension and longitude, and appulse—may occur hours or even days apart, and in different orders, depending on the directions in which the bodies are moving. Conjunction in r.a. can even happen without conjunction in longitude, or vice versa, if a planet happens to be near to where it reverses direction (the stationary moment at one end of its retrograde path).

Which kind of conjunction?

There's the question as to which sub-event is most important for observing. At appulse the bodies are closest to each other; but a telescope is aligned equatorially, so at conjunction in r.a. they cross the field of view simultaneously. I have cited the appulse, for instance in my *Astronomical Calendar*. It's the event that always occurs. But it's tricky in that it can occur too much: there can be an appulse without one or both of the conjunctions, for instance when a planet is near one end of its retrograde path.

In the list that follows, dates are for conjunctions in r.a. But in the yearly charts, lines connect Venus with other planets at appulses. If at one of the dates you draw a vertical line from Venus to the course of the other planet, that would show how the conjunction in r.a. differs.

In the list, "UT" is the hour in Universal Time. The symbols before each event are a quick guide to how spectacular and observable it is. "!" means that the separation of the two bodies is less than 1°. "*" means that the other planet is brighter than magnitude 2. The third symbol refers to the elongation, or angular distance from the Sun. "E" means it is greater 15° on the evening side; "m," it is greater than 15° on the morning side; "0," it is less than 15°.

(A third criterion might be freedom from moonlight.)

The best conjunctions are those marked "!*"; or, if you're not an early riser, "!*E." These close-bright-evening conjunctions are, in the timespan we've tabulated:

2020 May 22, Venus-Mercury
2021 Jul 13, Venus-Mars
2023 Jan 22, Venus-Saturn
2023 Mar 2, Venus-Jupiter
2027 Nov 25, Venus-Mars

—the last being the closest.

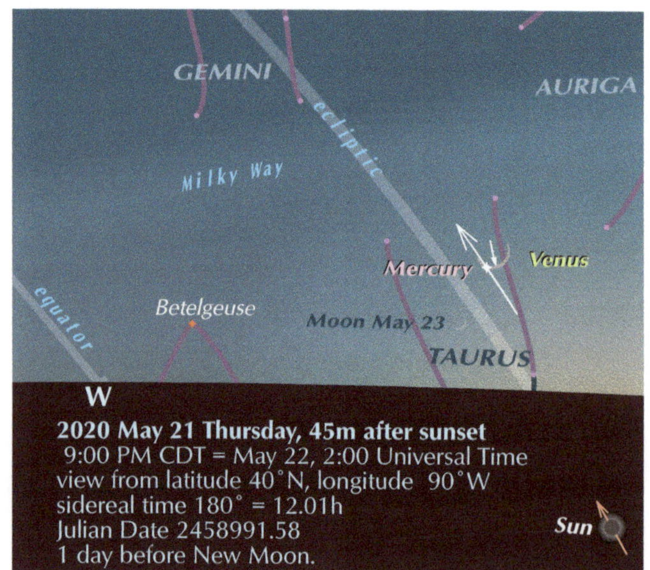

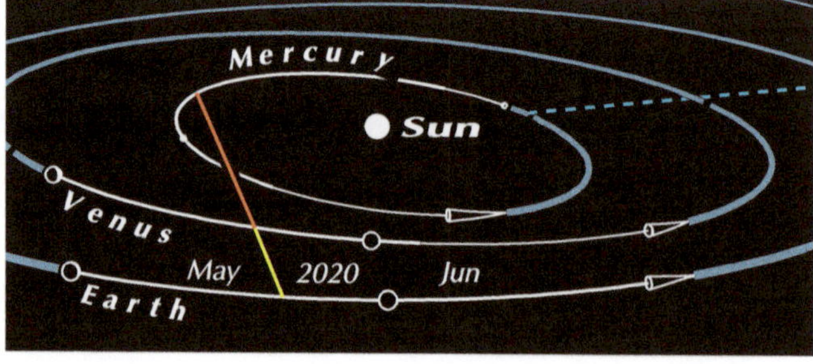

Mercury, 2020 May 22. Though Venus is 19° from the Sun and, for a northern latitude, about vertically above it at sunset, its crescent is very thin as it rushes toward the inferior conjunction of June 3.

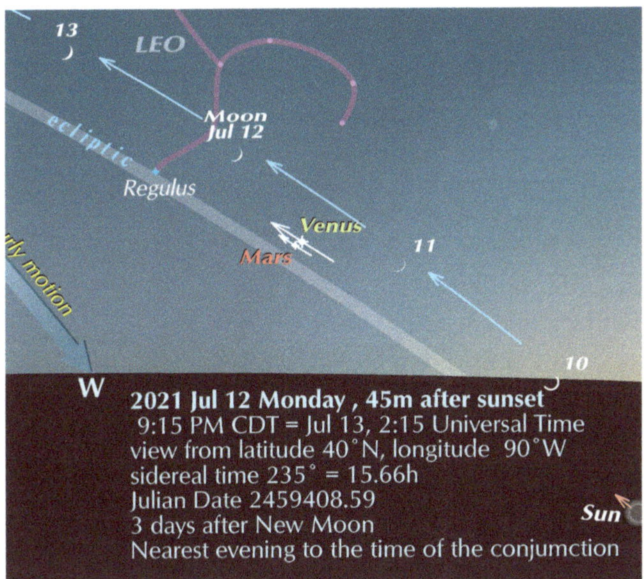

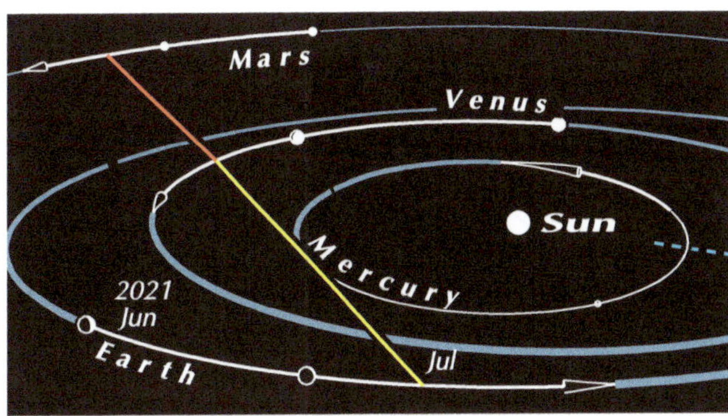

Mars, 2021 July 13. The northern summer angle of ecliptic to horizon puts the planets lower in the twilit sky. Venus is on the farther side of its orbit, moving out toward the easternmost elongation of Oct. 29.

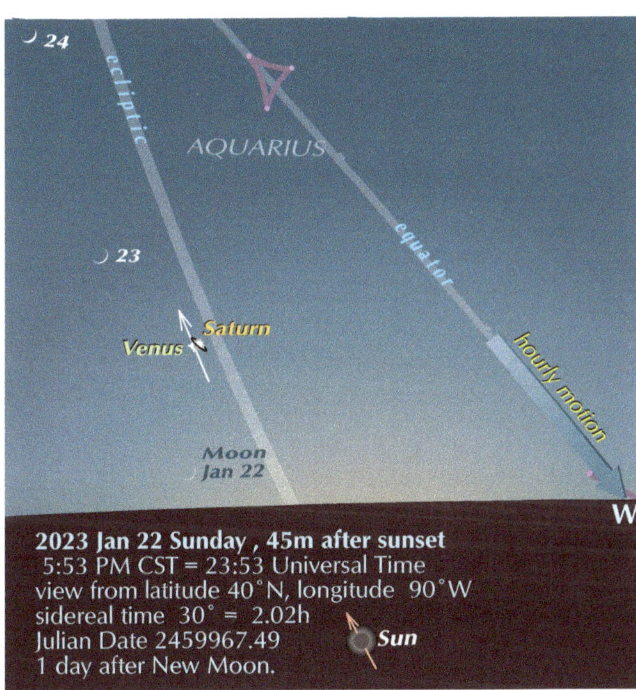

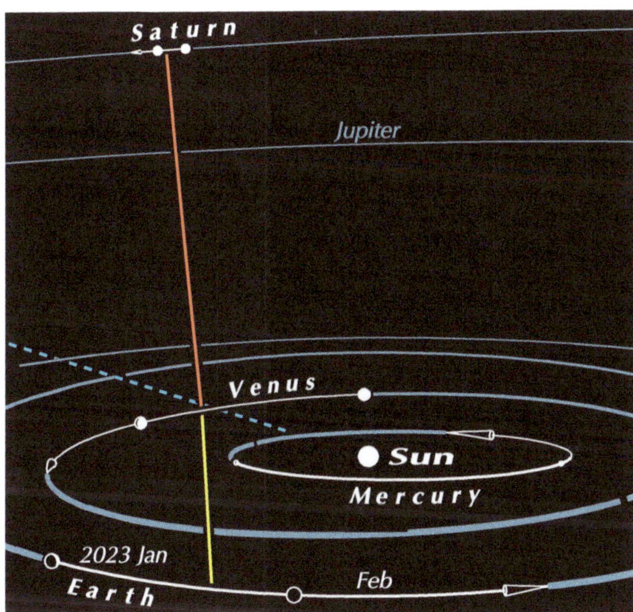

Saturn, 2023 Jan. 22. We see Venus overtake Saturn once in 378 days on average (Saturn's synodic period), but, as with Jupiter, the interval is irregular. The previous Venus-Saturn conjunction was 2021 Feb. 6, the next will be 2024 Mar. 2, both low in the morning sky.

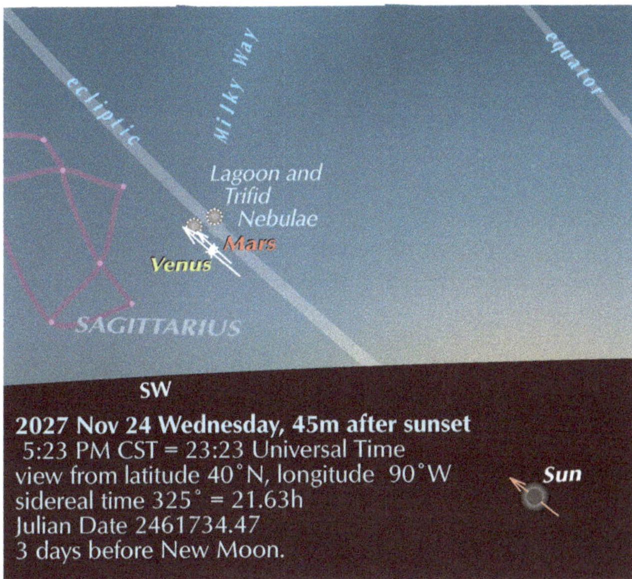

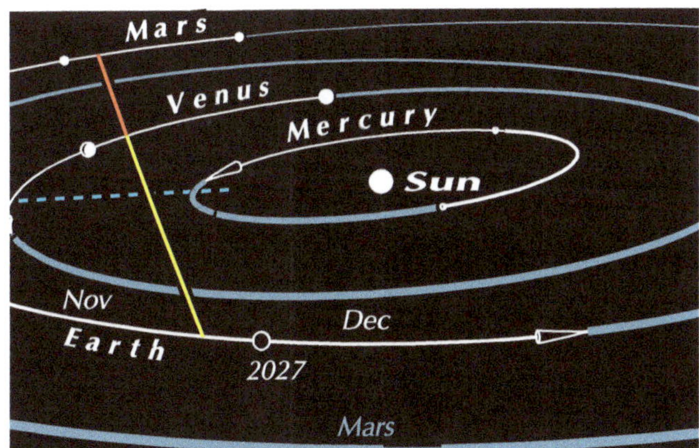

Mars, 2027 Nov. 25. Near the southernmost point of the ecliptic. About as favorable as 2021 July 13 (evening elongation slightly greater, separation between the planets slightly greater), and much better than 2026 Jan. 8 (very close to the Sun).

Venus-Jupiter conjunctions

The brightest and the second brightest planet, when they get together, can become a seeming double star that is noticeable in daylight and could even account for legends such as the Star of Bethlehem. There was a very close Venus-Jupiter conjunction in the June evening sky of 2 BC.

Venus meets Jupiter about yearly, to a first approximation. Jupiter, every 399 days–its synodic period–is behind the Sun; and Venus is always in the Sun's general neighborhood. But Venus may at that time be quite far east or west of the Sun; so its conjunctions with Jupiter can be as little as 10 or as much as 15 months apart. Sometimes there is a year with no Jupiter conjunction, preceded or followed by a year with two.

All Venus-Jupiter conjunctions are "bright," but about half are in the morning sky, and only a few meet our criterion of "close." The only close-bright evening ones between 1990 and 2050 are 1992 Aug. 23, 1999 Feb. 23, 2015 July 1, 2016 Aug. 27, 2023 Mar. 2, 2040 Sep. 1, and 2047 Mar. 8.

At most conjunctions, Venus in the foreground is simply overtaking Jupiter. But rather rarely there is triple conjunction. This happens around a time when Venus is passing in front of, and Jupiter behind, the Sun. Venus overtakes Jupiter in the evening sky, but then goes into its retrograde loop and falls back past Jupiter as both fall back past the Sun. Then Venus resumes forward motion and re-overtakes Jupiter in the morning sky.

There were triple conjunctions in 1991 and 2015. Those were 24 years apart–three eight-year

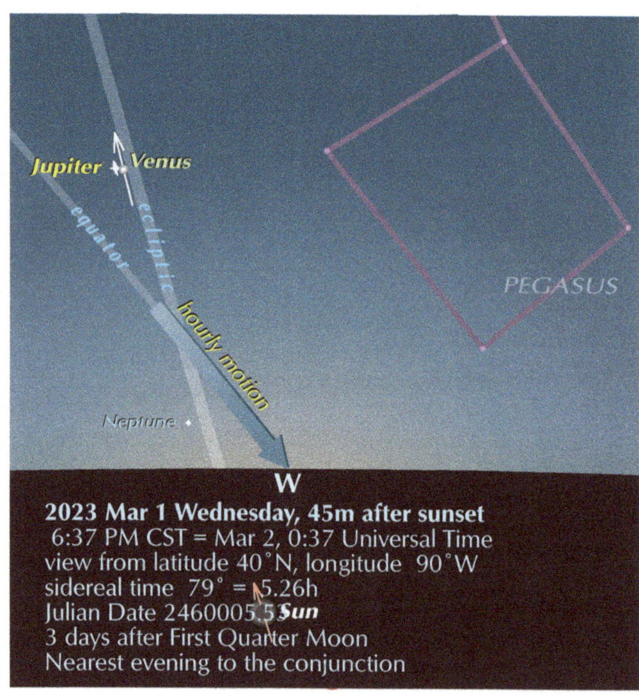

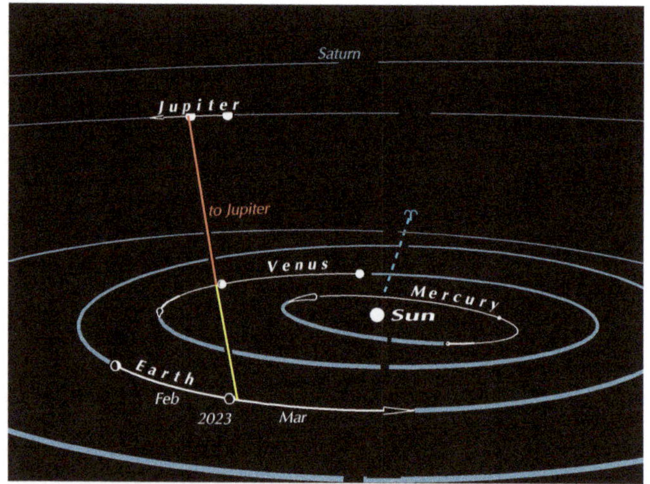

Jupiter, 2023 Mar. 1. The best Venus-Jupiter conjunction between 2016 Aug. 27 and 2040 Sep. 1.

```
Triple Venus-Jupiter conjunctions
1991    UT
Jun 17 23   *E   Venus 1.2° N of Jupiter; 45° from Sun in evening sky; mag -4.3, -1.8
Aug 23  7   *O   Venus 9.6° S of Jupiter; 8° from Sun in morning sky; mag -4.0, -1.7
Oct 17  4   *m   Venus 2.5° S of Jupiter; 45° from Sun in morning sky; mag -4.5, -1.8

2015
Jul  1 14  !*E   Venus 0.40° S of Jupiter; 42° from Sun in evening sky; mag -4.4, -1.8
Jul 31 20   *E   Venus 6.4° S of Jupiter; 22° from Sun in evening sky; mag -4.3, -1.7
Oct 26  8   *m   Venus 1.1° S of Jupiter; 46° from Sun in morning sky; mag -4.4, -1.8

2036
Mar 25  6   *E   Venus 4.3° N of Jupiter; 46° from Sun in evening sky; mag -4.4, -2.1
Jun  4  6  !*O   Venus 0.58° N of Jupiter; 8° from Sun in morning sky; mag -3.9, -2.0
Jul 23  4   *m   Venus 3.6° S of Jupiter; 45° from Sun in morning sky; mag -4.4, -2.1

2060
Apr  8  5   *E   Venus 5.5° N of Jupiter; 43° from Sun in evening sky; mag -4.5, -2.1
May 11 21   *E   Venus 4.9° N of Jupiter; 18° from Sun in evening sky; mag -4.2, -2.0
Aug  4  4   *m   Venus 2.9° S of Jupiter; 46° from Sun in morning sky; mag -4.3, -2.0
```

cycles. So they were similar, the conjunctions taking place against the background of Leo. But this is not a rule: the next triple conjunction will be 21 years later, in 2036. The situation rotates 90° backward: Jupiter, and the triple conjunction, are in Taurus. Then, another 24-year interval to a triple conjunction in 2060.

Conjunctions are not necessarily better for being part of a triple. The only close-bright-evening one in this timespan was 2015 July 1.

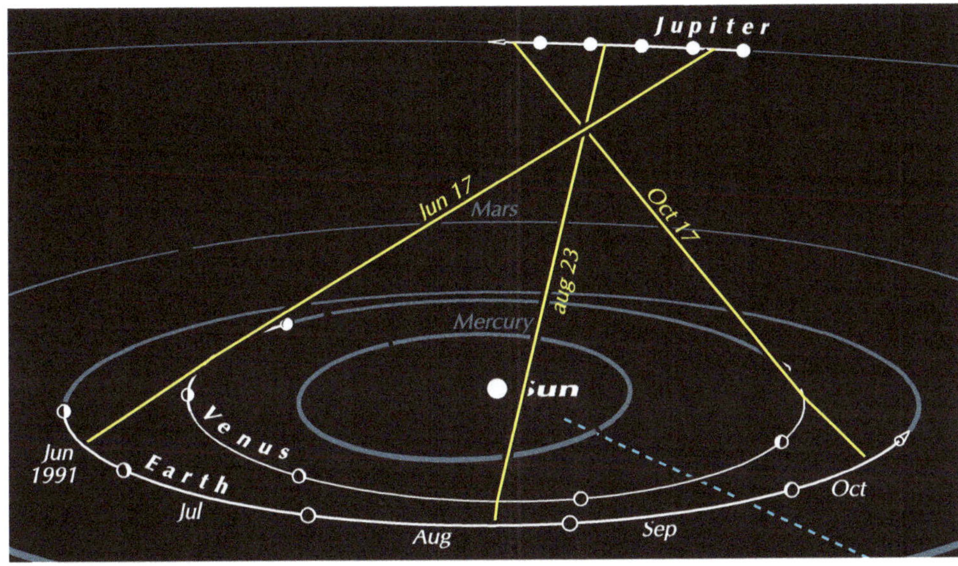

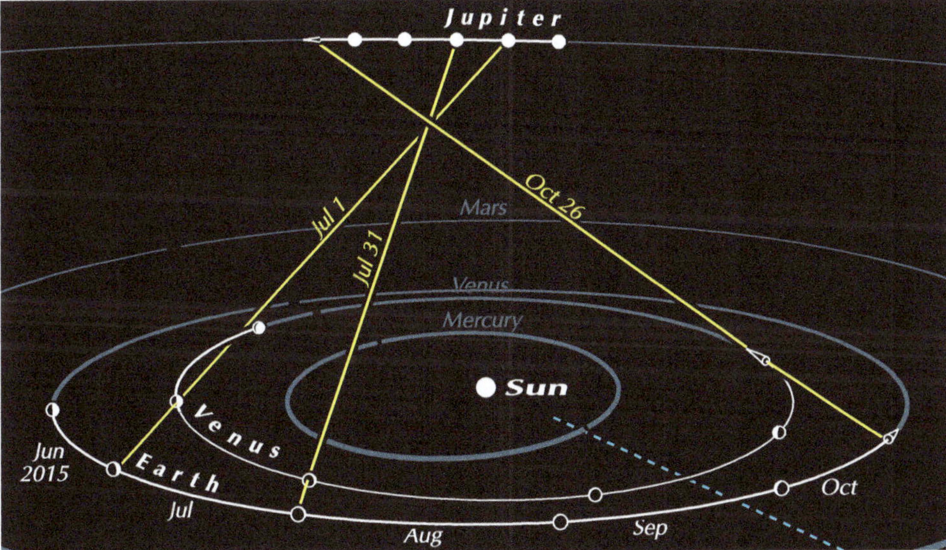

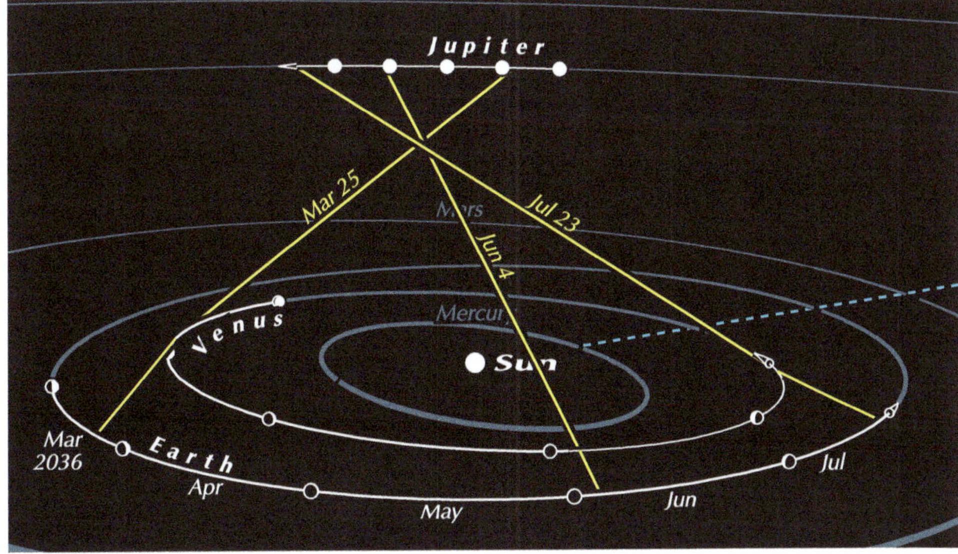

Triple conjunctions. While Venus whips around between us and the Sun, Jupiter plods along in the background. Great Jupiter is about four magnitudes fainter because of its distance. It is at about its most distant at these times, while Venus is about at its nearest.

The planets' courses are drawn for the five months that include the three dates. Their globes are drawn at the beginnings of the months; Earth and Venus are exaggerated 300 times in size, Jupiter 50, the Sun 5. The sightlines from Earth to Venus and from Venus to Jupiter bend slightly when Venus is above or below the ecliptic plane.

The dashed line from the Sun is the vernal equinox direction (longitude 0). The viewpoints are 6 astronomical units from the Sun, and 15° north of the ecliptic plane. In the 1991 and 2015 pictures, the view is from ecliptic longitude 330°; in the 2036 picture it has rotated 90°, to 240°.

The middle conjunction—the one that happens when Venus is retrograding—can be before or after Venus's inferior conjunction with the Sun. In 1991 and 2036 it is just after—8° into the morning sky. In 2036 and 2060, well before—still 22° and 18° out on the evening side.

Venus conjunctions with planets

2019	UT		
Jan 22	6	*m	Venus 2.4° N of Jupiter; 46° from Sun in morning sky; mag -4.3, -1.8
Feb 18	14	*m	Venus 1.1° N of Saturn; 43° from Sun in morning sky; mag -4.1, 0.7
Apr 10	4	! m	Venus 0.31° S of Neptune; 33° from Sun in morning sky; mag -3.9, 8.0
May 18	8	m	Venus 1.2° S of Uranus; 24° from Sun in morning sky; mag -3.9, 5.9
Jul 24	10	0	Venus 5.7° N of Mercury; 6° from Sun in morning sky; mag -3.9, 4.3
Aug 24	13	!*0	Venus 0.31° N of Mars; 3° from Sun in evening sky; mag -3.9, 1.8
Sep 13	22	!*0	Venus 0.34° N of Mercury; 8° from Sun in evening sky; mag -3.9, -0.9
Oct 30	8	*E	Venus 2.7° N of Mercury; 20° from Sun in evening sky; mag -3.9, 0.4
Nov 24	14	*E	Venus 1.4° S of Jupiter; 26° from Sun in evening sky; mag -3.9, -1.8
Dec 11	5	*E	Venus 1.8° S of Saturn; 30° from Sun in evening sky; mag -4.0, 0.6

2020	UT		
Jan 27	19	! E	Venus 0.07° S of Neptune; 40° from Sun in evening sky; mag -4.1, 7.9
Mar 9	15	E	Venus 2.4° N of Uranus; 45° from Sun in evening sky; mag -4.3, 5.9
May 22	8	!*E	Venus 0.90° N of Mercury; 19° from Sun in evening sky; mag -4.2, -0.6

2021	UT		
Feb 6	5	!*0	Venus 0.39° S of Saturn; 12° from Sun in morning sky; mag -3.9, 0.7
Feb 11	12	!*0	Venus 0.45° S of Jupiter; 11° from Sun in morning sky; mag -3.9, -2.0
Feb 12	17	0	Venus 4.8° S of Mercury; 10° from Sun in morning sky; mag -3.9, 3.0
Mar 14	1	! 0	Venus 0.40° S of Neptune; 3° from Sun in morning sky; mag -3.9, 8.0
Apr 22	23	! 0	Venus 0.25° S of Uranus; 7° from Sun in evening sky; mag -3.9, 5.9
Apr 26	9	*0	Venus 1.3° S of Mercury; 8° from Sun in evening sky; mag -3.9, -1.6
May 29	5	! E	Venus 0.42° N of Mercury; 17° from Sun in evening sky; mag -3.9, 2.3
Jul 13	7	!*E	Venus 0.49° N of Mars; 28° from Sun in evening sky; mag -3.9, 1.8
Dec 29	1	*E	Venus 4.2° N of Mercury; 17° from Sun in evening sky; mag -4.4, -0.7

2022	UT		
Feb 13	1	*m	Venus 6.6° N of Mars; 40° from Sun in morning sky; mag -4.6, 1.3
Mar 12	15	*m	Venus 4.0° N of Mars; 46° from Sun in morning sky; mag -4.5, 1.2
Mar 29	13	*m	Venus 2.2° N of Saturn; 46° from Sun in morning sky; mag -4.3, 0.9
Apr 27	19	! m	Venus 0.01° S of Neptune; 43° from Sun in morning sky; mag -4.1, 7.9
Apr 30	19	!*m	Venus 0.25° S of Jupiter; 43° from Sun in morning sky; mag -4.1, -2.1
Jun 11	13	m	Venus 1.6° S of Uranus; 34° from Sun in morning sky; mag -3.9, 5.8
Sep 26	1	0	Venus 3.8° N of Mercury; 7° from Sun in morning sky; mag -3.9, 3.9
Nov 22	17	*0	Venus 1.4° N of Mercury; 8° from Sun in evening sky; mag -3.9, -0.7
Dec 29	9	*E	Venus 1.4° S of Mercury; 17° from Sun in evening sky; mag -3.9, 0.4

2023	UT		
Jan 22	20	!*E	Venus 0.37° S of Saturn; 22° from Sun in evening sky; mag -3.9, 0.9
Feb 15	12	! E	Venus 0.01° S of Neptune; 28° from Sun in evening sky; mag -4.0, 8.0
Mar 2	11	!*E	Venus 0.54° N of Jupiter; 31° from Sun in evening sky; mag -4.0, -2.1
Mar 31	6	E	Venus 1.3° N of Uranus; 37° from Sun in evening sky; mag -4.0, 5.8
Jul 26	13	*E	Venus 5.3° S of Mercury; 25° from Sun in evening sky; mag -4.3, -0.1

2024	UT		
Feb 22	16	!*m	Venus 0.63° N of Mars; 26° from Sun in morning sky; mag -3.9, 1.3
Mar 22	2	!*m	Venus 0.35° N of Saturn; 19° from Sun in morning sky; mag -3.9, 1.1
Apr 3	11	! m	Venus 0.29° S of Neptune; 16° from Sun in morning sky; mag -3.9, 8.0
Apr 18	23	0	Venus 2.0° S of Mercury; 12° from Sun in morning sky; mag -3.9, 3.3
May 18	9	! 0	Venus 0.47° S of Uranus; 5° from Sun in morning sky; mag -3.9, 5.8
May 23	9	!*0	Venus 0.20° N of Jupiter; 3° from Sun in morning sky; mag -3.9, -2.0
Jun 17	13	!*0	Venus 0.88° S of Mercury; 4° from Sun in evening sky; mag -3.9, -1.9
Aug 6	15	*E	Venus 5.9° N of Mercury; 17° from Sun in evening sky; mag -3.9, 1.7

2025	UT		
Jan 20	5	*E	Venus 2.5° N of Saturn; 47° from Sun in evening sky; mag -4.5, 1.2
Feb 3	20	E	Venus 4.0° N of Neptune; 44° from Sun in evening sky; mag -4.6, 7.9

```
Mar  9 11    *E  Venus 6.3° N of Mercury; 22° from Sun in evening sky; mag -4.4, -0.1
Apr 29  2    *m  Venus 3.7° N of Saturn; 40° from Sun in morning sky; mag -4.5, 1.2
May  4  3     m  Venus 2.1° N of Neptune; 42° from Sun in morning sky; mag -4.5, 7.9
Jul  4  1     m  Venus 2.4° S of Uranus; 43° from Sun in morning sky; mag -4.1, 5.8
Aug 12  8   !*m  Venus 0.86° S of Jupiter; 36° from Sun in morning sky; mag -4.0, -1.9
Nov 25  5     0  Venus 1.1° S of Mercury; 10° from Sun in morning sky; mag -3.9, 2.1

2026   UT
Jan  8  4   !*0  Venus 0.17° N of Mars; 1° from Sun in evening sky; mag -3.9, 1.2
Jan 28 24   !*0  Venus 0.74° N of Mercury; 5° from Sun in evening sky; mag -3.9, -1.3
Feb 26 23    *0  Venus 4.7° S of Mercury; 12° from Sun in evening sky; mag -3.9, 1.2
Mar  7 12   ! 0  Venus 0.07° N of Neptune; 14° from Sun in evening sky; mag -3.9, 8.0
Mar  8 22    *0  Venus 1.0° N of Saturn; 15° from Sun in evening sky; mag -3.9, 1.0
Apr 24  5   ! E  Venus 0.78° N of Uranus; 26° from Sun in evening sky; mag -3.9, 5.8
Jun  9 12    *E  Venus 1.6° N of Jupiter; 37° from Sun in evening sky; mag -4.0, -1.9
Oct  5 15    *E  Venus 5.4° S of Mercury; 27° from Sun in evening sky; mag -4.5, -0.0

2027   UT
Apr 24  9   ! m  Venus 0.25° S of Neptune; 29° from Sun in morning sky; mag -3.9, 7.9
May  8  0   !*m  Venus 0.65° N of Saturn; 26° from Sun in morning sky; mag -3.9, 0.6
Jun 13 24   ! m  Venus 0.72° S of Uranus; 16° from Sun in morning sky; mag -3.9, 5.8
Jul  1 10     0  Venus 4.6° N of Mercury; 11° from Sun in morning sky; mag -3.9, 3.1
Aug 11  2   !*0  Venus 0.54° S of Mercury; 1° from Sun in evening sky; mag -3.9, -2.0
Aug 25 22   !*0  Venus 0.53° N of Jupiter; 4° from Sun in evening sky; mag -3.9, -1.7
Oct 10 13    *E  Venus 4.2° N of Mercury; 16° from Sun in evening sky; mag -3.9, 1.3
Nov 25  2   !*E  Venus 0.31° S of Mars; 27° from Sun in evening sky; mag -3.9, 1.3

2028   UT
Feb 11 24     E  Venus 1.3° N of Neptune; 43° from Sun in evening sky; mag -4.2, 7.9
Mar  1 12    *E  Venus 3.8° N of Saturn; 45° from Sun in evening sky; mag -4.2, 0.6
Apr 14  0     E  Venus 4.7° N of Uranus; 44° from Sun in evening sky; mag -4.5, 5.8
May 24  2     0  Venus 2.5° N of Mercury; 13° from Sun in evening sky; mag -4.1, 3.1
Jun  3  6   ! 0  Venus 0.44° N of Uranus; 3° from Sun in morning sky; mag -3.8, 5.8
Jun 14 16    *m  Venus 2.2° S of Mars; 20° from Sun in morning sky; mag -4.2, 1.4
Jul 16 14     m  Venus 4.4° S of Uranus; 43° from Sun in morning sky; mag -4.5, 5.8
Sep  8 22    *m  Venus 2.3° S of Mars; 44° from Sun in morning sky; mag -4.2, 1.6
Nov  9 19   !*m  Venus 0.66° N of Jupiter; 32° from Sun in morning sky; mag -4.0, -1.7

2029   UT
Jan 22 24    *0  Venus 3.7° S of Mercury; 15° from Sun in morning sky; mag -3.9, 1.7
Mar 27 18   !*0  Venus 0.26° S of Mercury; 2° from Sun in evening sky; mag -3.9, -1.9
Mar 27 22   ! 0  Venus 0.12° N of Neptune; 2° from Sun in evening sky; mag -3.9, 8.0
Apr 25  9    *0  Venus 1.7° N of Saturn; 8° from Sun in evening sky; mag -3.9, 0.3
May  5  3     0  Venus 1.9° S of Mercury; 11° from Sun in evening sky; mag -3.9, 3.2
May 19 11   ! 0  Venus 0.45° N of Uranus; 15° from Sun in evening sky; mag -3.9, 5.8
Sep  8  8    *E  Venus 1.9° S of Jupiter; 42° from Sun in evening sky; mag -4.1, -1.7
Nov 13  1    *E  Venus 2.5° S of Mars; 46° from Sun in evening sky; mag -4.5, 1.1
Nov 26  9    *E  Venus 2.0° S of Mars; 42° from Sun in evening sky; mag -4.6, 1.1

2030   UT
Jan 22 22    *m  Venus 5.5° N of Mercury; 25° from Sun in morning sky; mag -4.5, -0.1
May 12 21   ! m  Venus 0.32° S of Neptune; 40° from Sun in morning sky; mag -4.0, 7.9
Jun 25  1   !*m  Venus 0.32° N of Saturn; 31° from Sun in morning sky; mag -3.9, 0.2
Jul  9 12     m  Venus 1.1° S of Uranus; 27° from Sun in morning sky; mag -3.9, 5.7
Aug  6  0   !*m  Venus 0.71° S of Mars; 20° from Sun in morning sky; mag -3.9, 1.7
Sep  5  0     0  Venus 4.0° N of Mercury; 12° from Sun in morning sky; mag -3.9, 2.6
Oct  7 11   !*0  Venus 0.10° S of Mercury; 4° from Sun in morning sky; mag -3.9, -1.4
Nov 20 18   !*0  Venus 0.61° S of Jupiter; 8° from Sun in evening sky; mag -3.9, -1.7
Dec  9 24   !*0  Venus 0.71° S of Mercury; 13° from Sun in evening sky; mag -3.9, 1.6
```

Venus-Jupiter conjunctions

```
2019   UT
1990 Aug 12 23    !*m  Venus 0.04° N of Jupiter; 21° from Sun in morning sky; mag -3.9, -1.8

1991 Jun 17 23     *E  Venus 1.2° N of Jupiter; 45° from Sun in evening sky; mag -4.3, -1.8
     Aug 23  7     *0  Venus 9.6° S of Jupiter; 8° from Sun in morning sky; mag -4.0, -1.7
     Oct 17  4     *m  Venus 2.5° S of Jupiter; 45° from Sun in morning sky; mag -4.5, -1.8

1992 Aug 23  3    !*E  Venus 0.29° N of Jupiter; 19° from Sun in evening sky; mag -3.9, -1.7
1993 Nov  8 17    !*m  Venus 0.39° N of Jupiter; 17° from Sun in morning sky; mag -3.9, -1.7
1995 Jan 14  9     *m  Venus 2.8° N of Jupiter; 47° from Sun in morning sky; mag -4.4, -1.8
     Nov 19 12     *E  Venus 1.3° S of Jupiter; 23° from Sun in evening sky; mag -3.9, -1.8
1997 Feb  5 24    !*0  Venus 0.33° S of Jupiter; 14° from Sun in morning sky; mag -3.9, -1.9
1998 Apr 23  2    !*m  Venus 0.29° N of Jupiter; 45° from Sun in morning sky; mag -4.2, -2.1
1999 Feb 23 21    !*E  Venus 0.15° N of Jupiter; 28° from Sun in evening sky; mag -4.0, -2.1
2000 May 17 10    !*0  Venus 0.01° N of Jupiter; 7° from Sun in morning sky; mag -3.9, -2.0
2001 Aug  5 24     *m  Venus 1.2° S of Jupiter; 38° from Sun in morning sky; mag -4.0, -2.0
2002 Jun  3 18     *E  Venus 1.6° N of Jupiter; 34° from Sun in evening sky; mag -4.0, -1.9
2003 Aug 21  6    !*0  Venus 0.56° N of Jupiter; 2° from Sun in evening sky; mag -3.9, -1.7
2004 Nov  4 21    !*m  Venus 0.59° N of Jupiter; 35° from Sun in morning sky; mag -4.0, -1.7
2005 Sep  2 12     *E  Venus 1.4° S of Jupiter; 39° from Sun in evening sky; mag -4.0, -1.7
2006 Nov 15 23    !*0  Venus 0.44° S of Jupiter; 5° from Sun in evening sky; mag -3.9, -1.7
2008 Feb  1 13    !*m  Venus 0.59° N of Jupiter; 32° from Sun in morning sky; mag -4.0, -1.9
     Dec  1  1     *E  Venus 2.0° S of Jupiter; 43° from Sun in evening sky; mag -4.1, -2.0
2010 Feb 16 21    !*0  Venus 0.58° S of Jupiter; 9° from Sun in evening sky; mag -3.9, -2.0
2011 May 11  9    !*m  Venus 0.61° S of Jupiter; 26° from Sun in morning sky; mag -3.9, -2.1
2012 Mar 15 11     *E  Venus 3.3° N of Jupiter; 46° from Sun in evening sky; mag -4.3, -2.1
2013 May 28 21     *E  Venus 1.0° N of Jupiter; 16° from Sun in evening sky; mag -3.9, -1.9
2014 Aug 18  4    !*m  Venus 0.20° N of Jupiter; 18° from Sun in morning sky; mag -3.9, -1.8

2015 Jul  1 14    !*E  Venus 0.40° S of Jupiter; 42° from Sun in evening sky; mag -4.4, -1.8
     Jul 31 20     *E  Venus 6.4° S of Jupiter; 22° from Sun in evening sky; mag -4.3, -1.7
     Oct 26  8     *m  Venus 1.1° S of Jupiter; 46° from Sun in morning sky; mag -4.4, -1.8

2016 Aug 27 22    !*E  Venus 0.07° N of Jupiter; 22° from Sun in evening sky; mag -3.9, -1.7
2017 Nov 13  6    !*0  Venus 0.28° N of Jupiter; 14° from Sun in morning sky; mag -3.9, -1.7
```

License my roving hands, and let them goe
Before, behind, between, above, below.

> –John Donne (1571-1631), "Elegy XIX: To His Mistress Going to Bed." Donne was a priest, and in his lifetime only a book of his devotional poems was published. Most of his poems had to be published posthumously; in 1633 this one was refused a licence to be included in a collection of his Poems, but in 1654 it got into an anthology along with other authors, *The Harmony of the Muses.*

Let us roll all our strength and all
Our sweetness up into one ball,
And tear our pleasures with rough strife
Through the iron gates of life:
Thus, though we cannot make our sun
Stand still, yet we will make him run.

> –Last lines (41-46) of "To His Coy Mistress," Andrew Marvell, circa 1649; published posthumously, 1681

Year	Month	Day	Hour	Code	Event
2019	Jan	22	6	*m	Venus 2.4° N of Jupiter; 46° from Sun in morning sky; mag -4.3, -1.8
	Nov	24	14	*E	Venus 1.4° S of Jupiter; 26° from Sun in evening sky; mag -3.9, -1.8
2021	Feb	11	12	!*0	Venus 0.45° S of Jupiter; 11° from Sun in morning sky; mag -3.9, -2.0
2022	Apr	30	19	!*m	Venus 0.25° S of Jupiter; 43° from Sun in morning sky; mag -4.1, -2.1
2023	Mar	2	11	!*E	Venus 0.54° N of Jupiter; 31° from Sun in evening sky; mag -4.0, -2.1
2024	May	23	9	!*0	Venus 0.20° N of Jupiter; 3° from Sun in morning sky; mag -3.9, -2.0
2025	Aug	12	8	!*m	Venus 0.86° S of Jupiter; 36° from Sun in morning sky; mag -4.0, -1.9
2026	Jun	9	12	*E	Venus 1.6° N of Jupiter; 37° from Sun in evening sky; mag -4.0, -1.9
2027	Aug	25	22	!*0	Venus 0.53° N of Jupiter; 4° from Sun in evening sky; mag -3.9, -1.7
2028	Nov	9	19	!*m	Venus 0.66° N of Jupiter; 32° from Sun in morning sky; mag -4.0, -1.7
2029	Sep	8	8	*E	Venus 1.9° S of Jupiter; 42° from Sun in evening sky; mag -4.1, -1.7
2030	Nov	20	18	!*0	Venus 0.61° S of Jupiter; 8° from Sun in evening sky; mag -3.9, -1.7
2032	Feb	7	3	!*m	Venus 0.34° N of Jupiter; 29° from Sun in morning sky; mag -4.0, -1.9
	Dec	7	20	*E	Venus 1.9° S of Jupiter; 45° from Sun in evening sky; mag -4.2, -2.0
2034	Feb	22	19	!*0	Venus 0.45° S of Jupiter; 12° from Sun in evening sky; mag -3.9, -2.0
2035	May	17	16	!*m	Venus 0.53° S of Jupiter; 23° from Sun in morning sky; mag -3.9, -2.1
2036	Mar	25	6	*E	Venus 4.3° N of Jupiter; 46° from Sun in evening sky; mag -4.4, -2.1
	Jun	4	6	!*0	Venus 0.58° N of Jupiter; 8° from Sun in morning sky; mag -3.9, -2.0
	Jul	23	4	*m	Venus 3.6° S of Jupiter; 45° from Sun in morning sky; mag -4.4, -2.1
2037	Jun	3	7	*E	Venus 1.1° N of Jupiter; 19° from Sun in evening sky; mag -3.9, -1.9
2038	Aug	23	3	!*0	Venus 0.33° N of Jupiter; 15° from Sun in morning sky; mag -3.9, -1.8
2039	Nov	2	14	!*m	Venus 0.24° S of Jupiter; 46° from Sun in morning sky; mag -4.3, -1.8
2040	Sep	1	19	!*E	Venus 0.19° S of Jupiter; 25° from Sun in evening sky; mag -3.9, -1.7
2041	Nov	17	23	!*0	Venus 0.14° N of Jupiter; 11° from Sun in morning sky; mag -3.9, -1.7
2043	Jan	29	8	*m	Venus 2.0° N of Jupiter; 45° from Sun in morning sky; mag -4.2, -1.9
	Nov	29	21	*E	Venus 1.5° S of Jupiter; 29° from Sun in evening sky; mag -3.9, -1.9
2045	Feb	16	23	!*0	Venus 0.52° S of Jupiter; 7° from Sun in morning sky; mag -3.9, -2.0
2046	May	7	19	!*m	Venus 0.58° S of Jupiter; 40° from Sun in morning sky; mag -4.0, -2.1
2047	Mar	8	20	!*E	Venus 0.97° N of Jupiter; 34° from Sun in evening sky; mag -4.0, -2.1
2048	May	28	23	!*0	Venus 0.38° N of Jupiter; 0° from Sun in evening sky; mag -3.9, -2.0
2049	Aug	18	4	!*m	Venus 0.56° S of Jupiter; 33° from Sun in morning sky; mag -4.0, -1.9
2050	Jun	15	10	*E	Venus 1.5° N of Jupiter; 39° from Sun in evening sky; mag -4.0, -1.8

In the beginning, the Lord of Beings created men and women, and laid down a hundred thousand rules for their existence . . .

> Opening of the *Kama Sutra*, "Love Text." attributed to Vatsyâyana (perhaps around 300 BC), a Sanskrit treatise in prose and verse on how to find love and use it well; often regarded as a manual of sexual postures, but that is a small part of it.

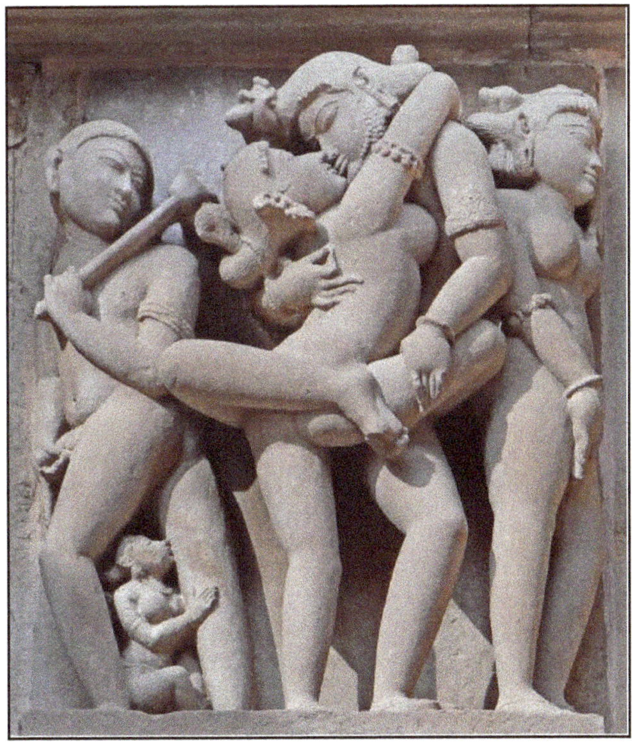

One of many erotic sculptures decorating the Khajuraho temples in India, from about 1000 AD.

Physical Venus

Venus, though Earth's nearest neighbor, was long a deep mystery. Mars and Mercury are naked, their surfaces open to view. Venus is completely clothed in a thick bright shawl of cloud.

Telescopes could not peer through that cloud; spectrometry could find out something about its composition. We did not know the planet's solid shape under the cloud, or even how or whether it rotates. More waited to be unveiled by the space probes to Venus began in the 1960s, and the first touchdowns on her in the 1970s. Even then, about all we knew for a while was that the planet has almost the same size and mass as Earth, and that it must have a dense, hot atmosphere. I remember an early speculation that the atmosphere, being so dense, would refract light all the way around in a circle, and you would see in the distance your own back!

What is now known is astonishing. Earth's "twin sister" differs from Earth in several profound ways. It's difficult to choose an order in which to describe them, because they are interrelated in a cat's-cradle of causal relations.

In brief, omitting some "almosts" and "roughlies":

Venus is perfectly spherical; rotates exactly in the plane of its orbit, but in a manner unlike any other planet; has a massive, oppressive, poisonous atmosphere; is not only hot under this atmosphere, but unvaryingly hot at all times and places—no variation between poles and equator, night and day, summer and winter; has a magnetic field, but not caused by an internal dynamo like Earth's; has no plate tectonic process like that which is so important to Earth; instead, suffers a vast convulsion of its crust at long intervals; could once have had habitable conditions, before its greenhouse effect became runaway.

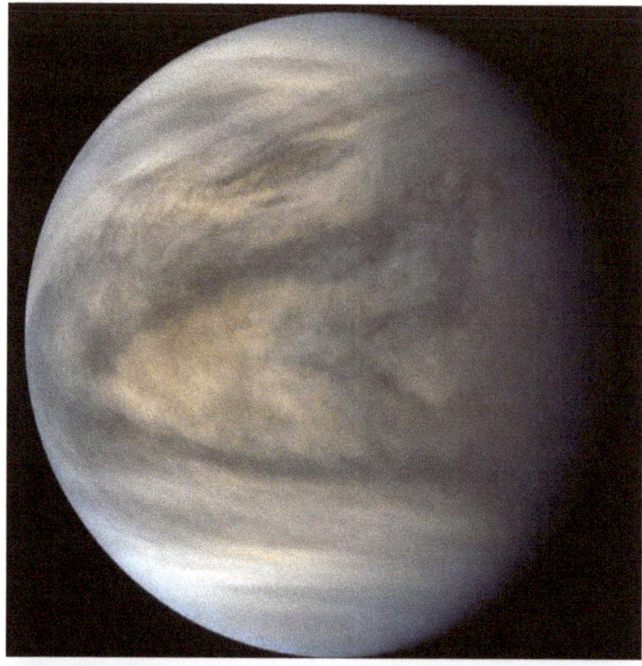

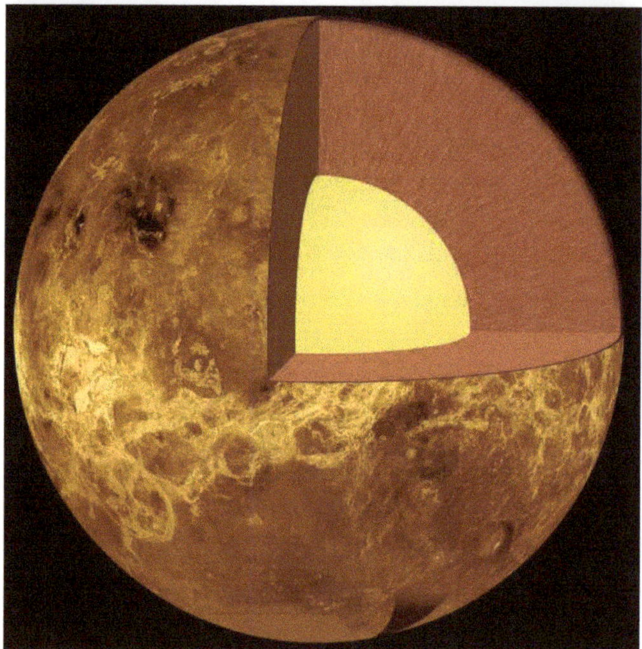

Core, mantle, and surface, in false color. NASA image.

Shape, attitude

Venus is exactly spherical, as far as can be measured. Its "flattening" is given as 0. Earth's is about 0.00335, meaning that its equatorial diameter is 42 km greater than its polar diameter. Rotation causes the equatorial bulge in planets such as Earth, and even more in the fast-spinning giant Saturn(flattening 0.098) Venus rotates very slowly.

And Venus is scarcely tilted. Its rotational axis is inclined only 2.64° to its orbital plane. Of the other planets, only Mercury and Jupiter have axial tilts about as small; more typical are Earth and Mars, both around 23°; Uranus is the opposite extreme, 98°.

Venus maintains this attitude throughout the ages. That is, having no Moon, it does not experience the slow change of axial tilt called precession, which causes Earth's spin axis to move in a 25,899-year circle among the stars.

We could say that among the bubbles that float around the solar-system pool, variously spinning and leaning, Venus is the prim little bubble that stays perfectly round and upright.

Rotation

Venus rotates extremely slowly: once in 243 Earth days. The rotation may be getting slower, rapidly: between 1990 and 2006 the Venus day grew 6.5 minutes longer.

And the rotation is backward: that is, westward, or clockwise as seen from the north. Bodies in the solar system generally both revolve and rotate eastward, or counter-clockwise, so this is called "direct" or "prograde" motion; Venus's rotation is retrograde.

The other exceptions among the major planets are Uranus, which rotates almost on its side, and Mercury, whose rotation period is exactly 2/3 of its year (in other words, it is tidally locked to the Sun, but not in the same way as the Moon's 1/1 locking to the Earth).

If you were on Venus and could see through the thick cloud—which you could not—you would see the stars rise in the west and take half of the 243 Earth days to cross the sky and set in the east. But, meanwhile, Venus is going around the Sun, which appears to move eastward past the stars, so that the time it takes to cross the Venus sky from west to east is shorter.

To sum up the numbers, in approximate Earth days:

117: the Venus day, as from a sunrise to a sunrise.

225: the Venus year, in which it goes around the Sun.

243: the rotation period, in which it spins once.

365: the Earth year, around the Sun.

584: the synodic period, as from one inferior conjunction to the next.

The shortest and longest of those reveal one further surprise: 584 / 117 is almost exactly 5. In other words, as Venus goes around from one inferior conjunction to the next, it rotates 5 times—and thus presents the same face to Earth! At each transit of Venus across the Sun, for instance, we have been looking at the same darkened hemisphere of Venus.

Does this mean that Venus, like the Moon, is locked in a tidal resonance with Earth? That has been a hypothesis, but has been rejected. The correspondence is not quite exact, and will evolve away, like the eight-year cycle.

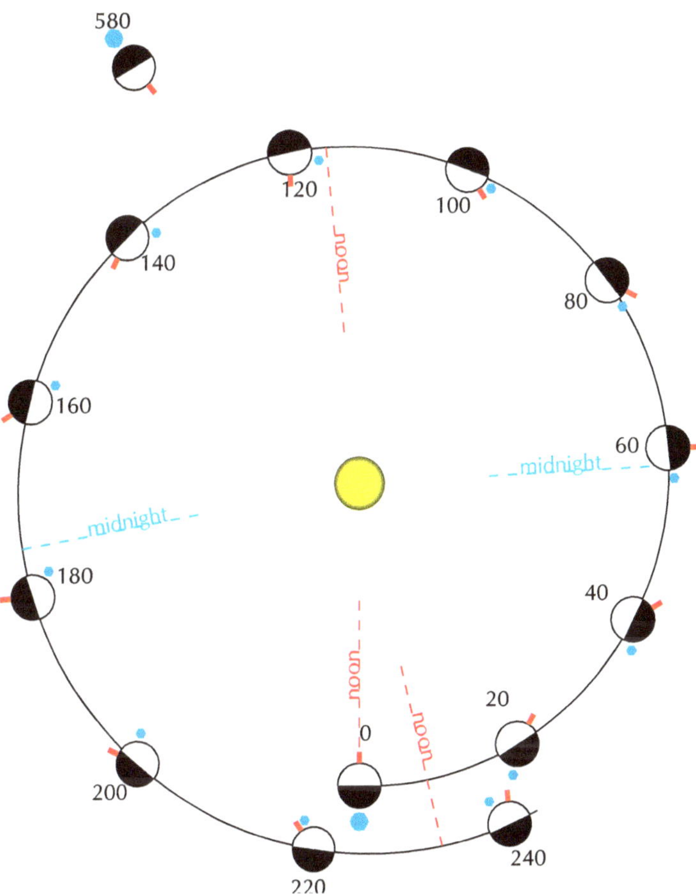

Rotation of Venus. The curve is a track of Venus, seen from the north, over a span of 243 Earth days. This is Venus's sidereal rotation period: the time during which it rotates once. That time is longer than Venus's orbital period of revolution around the Sun, 225 days, so we distort the curve outward to make the overlap clear. Venus is drawn (arbitrarily exaggerated in size) at intervals of 20 Earth days. The red sticks represent a single location on Venus's surface. You can think of them as a Venusian giant, very hot! At time 0 of the Venusian year shown, it is midday for the Venusian: he has the Sun at his zenith. As Venus rotates slowly *backward*, the Venusian is carried eastward (instead of westward as on Earth). After about 29 Earth days, the Venusian sees the Sun setting, on his *eastern* horizon. By about the 58th Earth day, the Venusian is facing outward: it is his midnight. And by Earth day 117 he is again facing the Sun. One Venus day has been completed.

Blue dots are off the side of Venus that faces Earth. You can see that Earth is over Venus's midnight meridian at day 0 and again at day 584—which could be the dates of two successive inferior conjunctions.

Surface

Venus has fewer impact craters than Earth. None are very small: space projectiles below a certain size cannot get down through the thick atmosphere without burning up. And none of the craters are very old, showing that the surface is younger.

Craters and rocks look sharp, nearly pristine. They are not eroded by bombardment from space particles as on the Moon or by water as on the Earth.

There are two highland areas: Ishtar Terra in the northern hemisphere and Aphrodite Terra in the southern, respectively as large as Australia and South America. In Ishtar Terra are the highest mountains, Maxwell Montes, reaching 17 km above the average surface. The remaining 80 percent of the surface is plains, some smooth, but mostly wrinkled with ridges.

There are many more volcanoes than on Earth. Earth's only volcano group over 100 km wide is the Big Island of Hawaii; Venus has 167 this size. Volcanoes have not been seen erupting, but there is much evidence that they do. They have pumped sulfur into the atmosphere, and coated the surface.

Some peculiar volcanic features are called "farra" because they are shaped like pancakes; others are "novae" and "arachnoids" because the cracks in them are reminiscent of stars or spiderwebs.

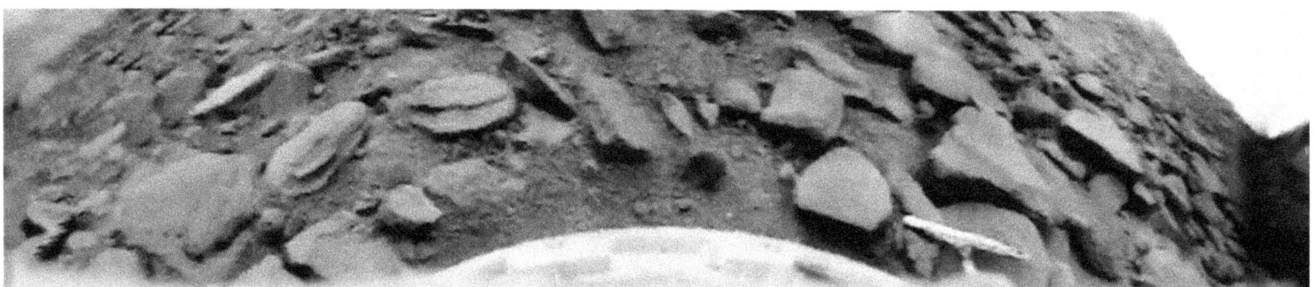

First photo of the surface, taken by the Soviet space probe Venera 9 in 1975.

Temperature

Venus is indeed hot. Its surface is at about 464° Celsius (867° Fahrenheit).

The temperature is mainly owed not to incoming sunlight, but to heat rising from the core. The planets formed by condensation out of the solar nebula, yielding high pressure at the center and heat that has to rise. In Venus's case the heat is trapped: greatly slowed from escaping by the greenhouse gases of the atmosphere.

This is obvious from the contrast with bare Mercury, which is almost twice as near to the Sun, yet whose temperature peaks at only 427° C on its Sun-baked side (and goes down to -173° C on its night side).

The heat is almost unvarying. The north and south poles are as hot as the equator. The night side remains as hot as the side facing the Sun, even though each night lasts as long as 58 Earth nights. This is because the Sun's radiation only slowly percolates down to the lower atmosphere, where the sluggish winds only slowly push it along.

Only height brings some cooling. The tops of the highest mountains are at a temperature of about 380°C (715°F).

Because Venus has no axial tilt, it has no seasons. Temperature scarcely varies as it moves around its orbit.

> A savage place! as holy and enchanted
> As e'er beneath a waning moon was haunted
> By woman wailing for her demon lover!
>
> –Coleridge, "Kubla Khan" (1797)

Atmosphere, clouds, wind, water

Venus has 93 times as much atmosphere as Earth, by mass. The density and pressure at the solid surface are correspondingly greater. It's like being half a mile down in Earth's ocean.

The atmosphere is more than 96 percent carbon dioxide, concentrated in the lowest part. Most of the rest is nitrogen. Higher are other layers; one consists of sulfuric acid, another of ozone. As on Earth, there is a very high ionosphere, of charged particles,

Water vapor, which forms about one percent of Earth's atmosphere, was probably "eroded" out of Venus's upper atmosphere by the solar wind, in the first billion years.

The clouds may produce lightning storms, but the evidence is disputed.

Winds in the dense lower atmosphere can move only at slow walking speeds. They are elephant winds compared with our bird winds! Or they are like vehicles in low gear. Because of their density, they can push dust and stones along.

But winds on the cloud tops can move at 300 kilometers an hour, or 60 times faster than the slow rotation speed of the surface under them, whereas Earth's fastest winds move at only a small fraction of the rotation speed.

Magnetism

Venus has a magnetic field, much weaker than Earth's. It is created not by a core dynamo, as in Earth's core, but by the interaction of the solar wind with the top of the atmosphere, the ionosphere.

Satellites?

Venus has no permanent satellite like Earth's Moon. Giovanni Cassini in 1672 thought he discovered one. It was named Neith, for an Egyptian goddess, and many later sightings of it were claimed; they were probably background stars. Small bodies from time to time become temporary satellites; and there are several that, like Jupiter's Trojan asteroids, cluster at 60° ahead and behind in Venus's orbit.

My luve is like a red, red rose
That's newly sprung in June,
My luve is like the melodie
That's sweetly play'd in tune.
As fair art thou, my bonnie lass,
So deep in luve am I,
And I will luve thee still, my dear,
Till a' the seas gang dry.

Till a' the seas gang dry, my dear,
And the rocks melt wi' the sun,
And I will luve thee still, my dear,
While the sands o' life shall run.
And fare thee weel, my only luve,
And fare thee weel, a while,
And I will come again, my Luve,
Tho' it were ten thousand mile.

–Robert h worked to preserve Scottish folksongs, and published in 1784 a collection, including "Auld Lang Syne." He had heard "Red, Red Rose" sung "by a country girl.".

Interior

Being about Earth-sized, Venus probably has similar structure: crust, mantle, and partly liquid core.

But they do not show convection—the cyclic rising of hot lower fluid and sinking of cooler, as in boiling water. Presumably the material is too rigid, and lacks lubrication by water. On Earth, convection allows the rotating core to be a dynamo, creating the magnetic field. And in the mantle and crust, it causes tectonics: the process by which continental plates are extruded, drift, and are subducted at their edges,

Upheaval

Heat rises; bodies are always getting rid of their heat. Heat cannot dissipate from Venus by the continual process of convection and plate tectonics. Instead (it is thought), it does so by periodic global convulsions.

The heat in the mantle, unable to escape through the crust, increases relentlessly, until it reaches a critical level. Then the crust gets broken up, presumably by a fever of quakes and eruptions. It sinks in fragments, is entirely recycled.

The last of these reshufflings happened about 600 million years ago, and took perhaps 100 million years. That is why the surface is "young": there are no craters older.

Venus at 2020 June 1, 12 UT, only 54 hours before inferior conjunction. Photo by James Weightman at Cirencester, England. The unfocused shadow at right is his house, blocking out the Sun. I hope you can see, at center, the extremely slender crescent. Venus is less than 4° northeast of the Sun; its disk is 58 seconds wide, but less than 1% sunlit; its brightness is reduced to magnitude -3.8 (still far brighter than any star). The specks, some brighter than Venus, are not the background stars in Taurus but sunlit motes of dust, in whizzing motion. (Some perhaps carrying coronavirus. In micrometers or millionths of a meter, dust is up to 500 wide; the virus, 0.125. If the virus were a tennis ball, a dust speck might be 300 yards across.) At lower left is a simultaneous image from SOHO (the Solar and Heliospheric Observatory satellite), which masks the Sun to reveal objects near it, such as Sun-grazing comets. The Sun, if not masked, would be about 2,000,000,000 times brighter than Venus.

Life?

It's possible that the early Venus had conditions closer to those that on Earth allowed molecules to assemble into forms that could self-replicate. But this would have been shut off by increasing heat, and the loss of water.

If Venus has any microbial life now, it would have to be in the cooler upper atmosphere. And, there, it would have to tolerate sulfuric acid. The rest of the planet is above the temperature needed for sterilizing.

Earth's greenhouse effect (stopping of heat from escaping into space), which is driving the average surface temperature above what it had been for thousands of years, is dwarfed by Venus's perpetual greenhouse condition. But we would be wise to learn from it. When feedback loops start (such as the shrinking of Arctic ice, which exposes more dark-colored water, which absorbs heat instead of reflecting it; or the melting of permafrost, which releases methane, which fountains into the air and is 25 times more efficient as a greenhouse gas than carbon dioxide), we cannot be sure when they will stop. Best to do all possible to stop them now.

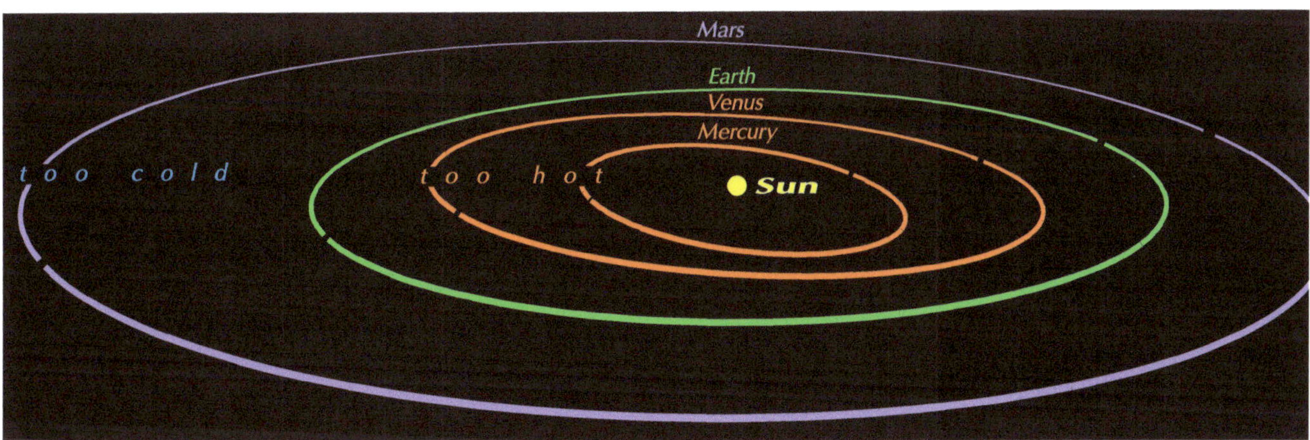

Orbits, seen from a viewpoint 6 astronomical units from the sun and 15° north of the ecliptic plane, at longitude 246° (to show the eccentricity of the Mars orbit). Venus and Mars orbit at about 0.7 and 1.5 a.u. And they, especially Venus, are similar in size and solid make-up to Earth. The band between too hot and too cold is narrow.

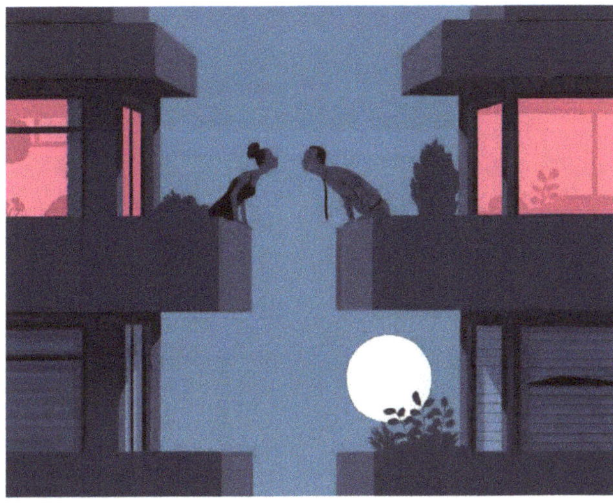

Love in the time of coronavirus. Cartoon by Paul Blow in *The Observer*, 2020 April 12.

Space exploration to Venus

There have been (as of 2020) over 40 space missions or attempted missions to Venus. They were operated by the Soviet Union, the U.S.A.'s NASA (National Aeronautic and Space Agency), JAXA (Japan Aerospace Exploration Agency), and E.S.A. (the European Space Agency). Fifteen failed, at launch or at later stages, because of failure of some mechanical part or because the craft ceased to be controlled from the ground or to send back data.

Typically a spacecraft is lifted on a rocket into geocentric orbit–around the Earth–and then separates from it, using its fuel to accelerate into a heliocentric orbit–around the Sun–so as to intersect the orbit of another planet. Sometimes the aim was flyhy (ungainly but unavoidable word). Sometimes this was a stage in a journey to further planets, or asteroids or comets. Or the aim was descent through Venus's thick atmosphere and impact on the surface, either of which would destroy the craft but result in information. Or the aim was to reach the planet's gravitational field and be inserted into orbit around it, from which a module could separate and descend into the atmosphere or to the surface.

We'll list the missions by launch date.

1961 Feb. 4; Soviet; **Tyazhely Sputnik**, meaning in Russian "heavenly traveling companion" (from the word *put*, "journey"), known while in development as Venera 1VA No. 1, and in the West as Sputnik 7. It failed to get out of low Earth orbit (called LEO by spaceniks), and the Soviet government maintained that it was only a test.

1961 Feb. 11; Soviet; **Venera 1**, or Sputnik 8. After a journey of two months it flew by Venus at a distance of less than 100,000 kilometers; but radio communication with it had been lost.

1962 July 22; NASA; **Mariner 1**. Control of it failed and it had to be destroyed less than five minutes after launch.

1962 Aug. 25; Soviet; **2MV-1 No.1**, or Sputnik 19. It failed to get out of low Earth orbit, and plunged back into the atmosphere a few days after launch.

1962 Aug. 27; NASA; **Mariner 2**. The first successful probe to any planet. It flew by Venus at a distance of 34,773 kilometers on Dec. 14, taking measurements of the cool cloud tops and hot surface.

1962 Sep. 1; Soviet; **2MV-1 No.2**. A fuel valve didn't open, so it failed to reach low Earth orbit.

1962 Sep. 12; Soviet; 2MV-2 No.1. Another

Mariner 1

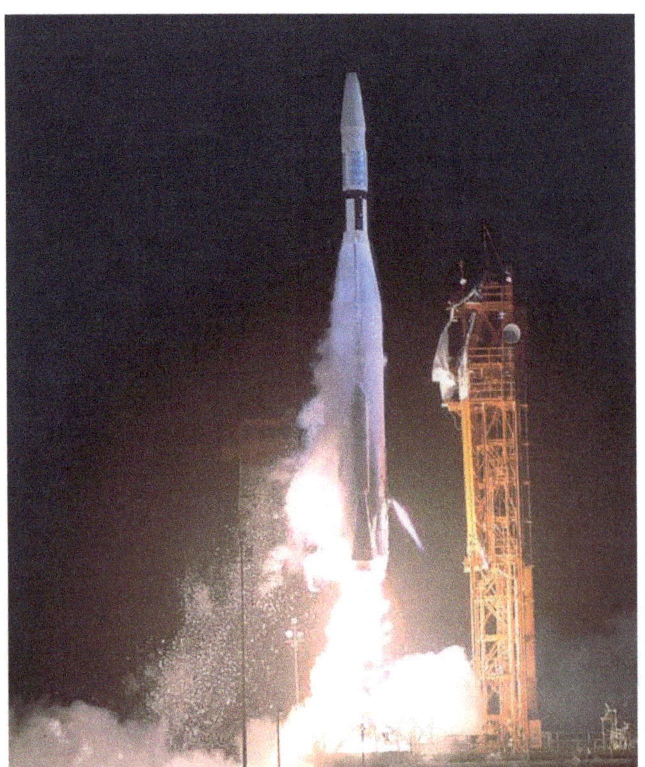

Mariner 2

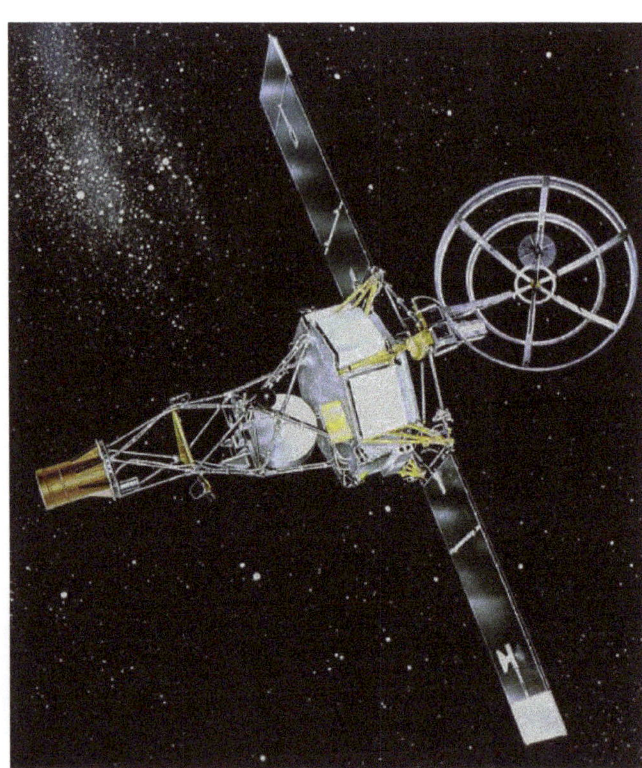

technical problem, within seconds.

1964 Feb. 19; Soviet; **3MV-1 No.2**. Another technical mishap at launch.

1964 Mar. 27; Soviet; **Kosmos 27.** The name "Kosmos" was given to craft that remained in Earth orbit; then they could be renamed if they successfully got beyond it. This one failed to do so.

1964 Apr. 2; Soviet; **Zond 1**. Flew past Venus on July 14, but had lost communication.

1965 Nov. 12; Soviet; **Venera 2**. Flew past Venus on 1966 Feb. 27, but communication had

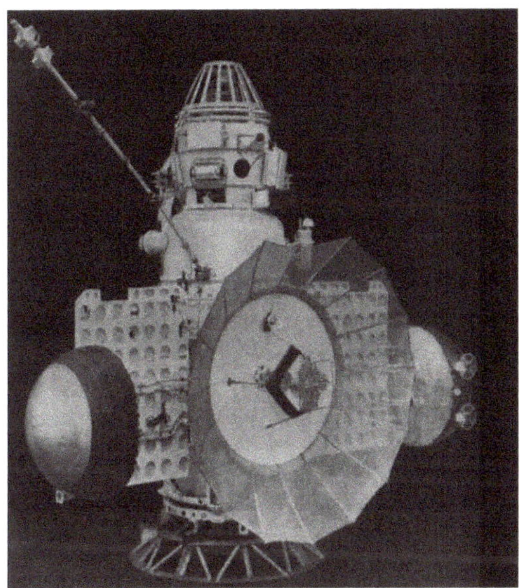

Zond 1

been lost.

1965 Nov. 16; Soviet; **Venera 3**. The first craft to hit Venus. Its first flyby was too wide; a course correction was made on Dec. 26, On 1966 Feb. 15, entry into Venus's atmosphere burned it up before it could send any data.

1965 Nov. 23; Soviet; **Kosmos 96**. A combustion chamber exploded; failed to reach low Earth orbit.

1967 June 12; Soviet; **Venera 4**. The breakthrough: the first to survive as it descended through the atmosphere of Venus, on Oct.18. While doing so, it transmitted data that revealed the types of gas, the lack of water, the weak magnetic field, the density, pressure, and heat. Its experience allowed later Veneras to be designed so that they could land usefully on the surface.

1967 June 14; NASA; **Mariner 5**. Flew by Venus on Oct. 19 at a height of 3,990 km, with more sensitive instruments than its predecessors for

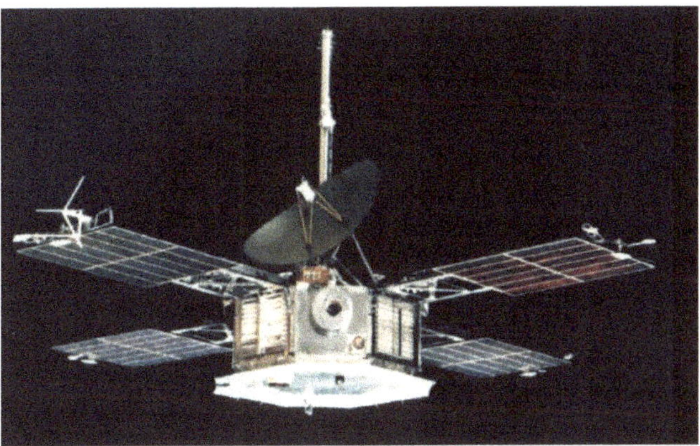

Mariner 5

analysing the atmosphere and surrounding space.

1967 June 17; Soviet; **Kosmos 167**. Failed to reach low Earth orbit.

1969 Jan. 5 and 10; Soviet; **Venera 5 and 6**. They arrived at Venus on May 16 and 17, and each dropped, from distances of 27,000 and 25,000 km, a 405 kilogram capsule containing instruments (and, in both capsules) the Soviet coat of arms and a bas relief of Lenin). The capsules hit the night side of Venus's atmosphere and were then slowed by parachutes, whose size was tuned, based on experience with Venera 4, to allow the capsules to get as far down as possible before succumbing to the heat. The instruments sent data every 45 seconds, for, respectively, 53 and 51 minutes.

1970 Aug. 17; Soviet; **Venera 7**. The first soft landing on Venus, Dec. 15. But it rolled over and could transmit little of its data.

1970 Aug. 22; Soviet; **Kosmos 359**. This was another Venera, designed in the same way as its predecessor and intended to land five days after it on Venus. But a malfunction sent it into an el-

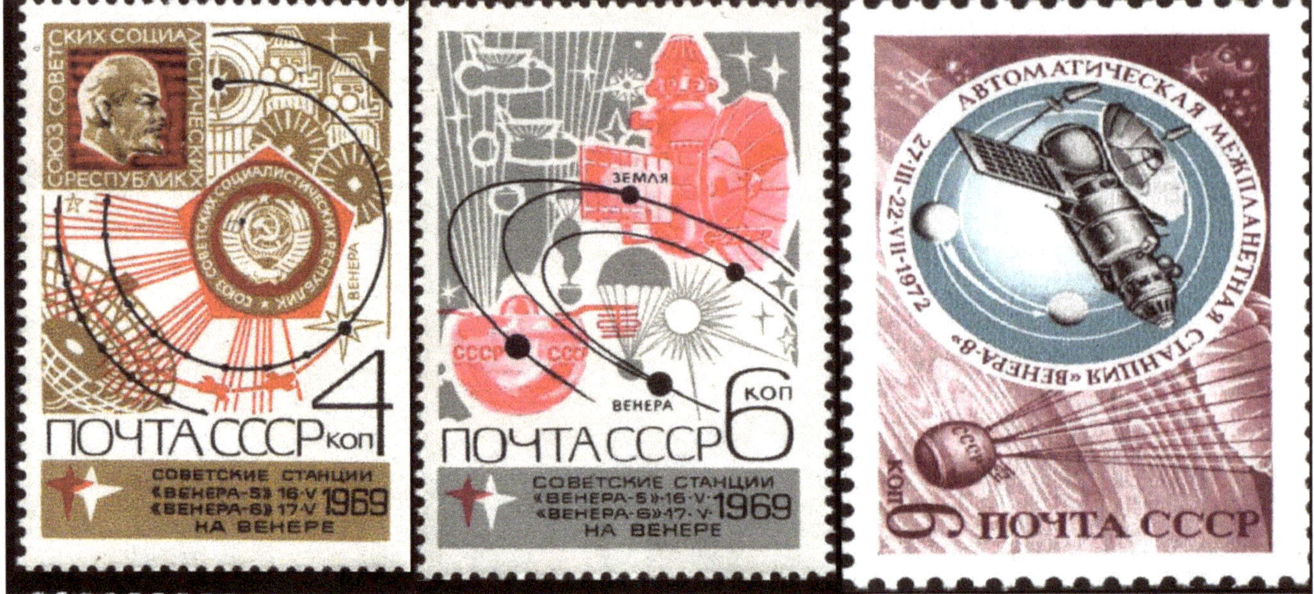

Stamps commemorating Veneras 5, 6, and 8

liptical orbit around Earth, until after 410 days it re-entered our atmosphere and burned up. So it kept the Kosmos name in the hope of concealing the failure.

1972 Mar. 27; Soviet; **Venera 8**. Landed on Venus July 19, the first fully successful landing on another planet. It settled on the surface and transmitted data until July 22.

1972 Mar. 31; Soviet; **Kosmos 482**. Never got out of low Earth orbit, Its landing module, designed to withstand 100 times Earth's atmospheric pressure, is likely to crash to Earth's surface between 2023 and 2025.

1973 Nov. 3; NASA; **Mariner 10**. First craft to visit more than one planet, and to use the gravity assist or "slingshot" method of bending the path to reach a new destination. It was also the first to use the pressure of the Sun's radiation to adjust its course. It flew by Venus on 1974 Feb. 4, was thus redirected to pass Mercury on March 29. Then it went once around the Sun while Mercury did so twice, and flew by Mercury a second time on Sep. 21; and made a third and closest Mercury encounter on 1975 March 16. Since then, without radio communication, it has presumably still been orbiting the Sun.

1975 June 8; Soviet; **Venera 9**. It went into orbit around Venus—the first to do so—from Oct. 20 to Dec. 25. Its lander fell from Oct. 20 to 22, protected at first by a spherical shell, with explosive bolts, which opened it to release a three-dome parachute. The lander touched down at a speed of about 8 meters a second, near the middle of the sunlit side of Venus, on a slope that was guessed to be the side of the rift valley Aikhylu Chasma. It was the first to return images (black-and-white) of the surface, showing rocks and soil. The images spanned 180° instead of 360° because one of the two cameras failed to open.

1975 June 14; Soviet; **Venera 10**. Entered orbit on Oct. 23. The lander fell, from Oct. 23 to 25; besides the shell and parachutes, it was protected by pre-cooling, a system of circulating fluids, a disk-shaped drag brake, and a doughnut-shaped compressible metal landing cushion. It landed 2,200 km from Venera 9, near the border between the areas called Beta Regio and Hyndla Regio, and was able to operate for 65 minutes, measuring temperature, pressure, density, and the wind speed of only 3.5 km per hour.

1978 Sep. 9; Soviet; **Venera 11**. The lander descended from Dec. 23 to 25. It had protection like its predecessors, and relayed information for 95 minutes until the mother craft, continuing on its orbit around the Sun, passed out of range. The cameras were now color instead of black-and-white; but both failed to open.

1978 Sep. 14; Soviet; **Venera 12**. The lander descended, with similar protection, from Dec. 19 to 21, and transmitted data for 110 minutes until out of range. Among its findings about the atmosphere were evidence for lightning and for some carbon monoxide as well as dioxide. Its color cameras, like those of Venera 11, failed to open.

1978 May 20; NASA; **Pioneer Venus Orbiter**, or Pioneer Venus 1, or Pioneer 12. On Dec. 4 it entered an elliptical Venus orbit, which was several times adjusted so as, for instance, to see the otherwise unobservable south pole. Besides measurements of Venus, it took a look far outward, at the time of year when this could not be done from Earth, at Halley's Comet, now at an enormous distance of 35 astronomical units; some of its findings were later confirmed at the comet's 1986 perihelion. The Pioneer kept sending data until 1992 Oct. 8. Then corrections ceased being made to its orbit, which decayed, so that the craft disintegrated into the atmosphere on Oct. 22.

1978 Aug. 8; NASA; **Pioneeer Venus Multiple**, or Pioneer Venus 2, or Pioneer 13. It released a large probe on Nov. 16 and three smaller probes on Nov. 20. Then on Dec. 9 all entered Venus's atmosphere, the bus last and at a shallow angle, putting it in orbit; it continued sending signals for only a minute. The large probe, only, had a parachute. The smaller ones were named North, Night, and Day, for the parts of the planet they were aimed at. The probes did not have cameras, and were designed to keep making measurements of the atmosphere until they hit the surface. But Day survived the impact and continued sending information for an hour.

1981 Oct. 30 and Nov. 4; Soviet; **Venera 13 and 14**. Identical craft, built to take advantage of a 1981 Venus launch opportunity. They reached Venus on 1982 March 1 and 5, and their landers, slowed by parachutes that opened 50 km above the surface, settled in the plain called Phoebe Regio. The cameras took panoramic pictures, and there were instruments to analyse the soil by probing and sampling. Venera 14, however, measured the compressibility of a camera lens cap, which had got underneath it after popping off. Two scientists, Leonid Ksanfomaliti (who had contributed to the mission) and Stan Karaszewski, published an article claiming that Venera 13 photographs showed signs of life: shapes that moved, a "disk," a "black flap," and a "scorpion." Others refuted this; the disk that "moved" was two parts of the camera lens cap.

1983 June 2 and 7; Soviet; **Venera 15 and 16**. Identical orbiters, to map the surface of Venus, using radar. On Oct. 10 and 11, they entered strongly elliptical orbits that were about 24 hours long, nearly polar, and 4° away from to each other so as to re-image areas if necessary. They operated till 1984 July. Because the southern ends of their orbits were 6,400 km farther out than the northern, they could map only the northern quarter of Venus, down to about latitude 30° north.

1984 Dec. 15 and 24; Soviet; **Vega 1 and 2**. "VeGa" did not refer to the star, or the Spanish word for a meadow, but was an acronym of "Venus" and "Halley" (Russian turns the foreign consonant h into g). 1P Halley, most famous of comets, was approaching. (Its designation means that it is the first in the list of periodic comets.) The U.S. proposed a Halley mission, but canceled it in 1981. So the Soviets repurposed what would have been the next two craft in the Venera program, with cooperation from France, West Germany, Austria, East Germany, Poland, Czechoslovakia, Hungary, and Bulgaria. The two Vegas

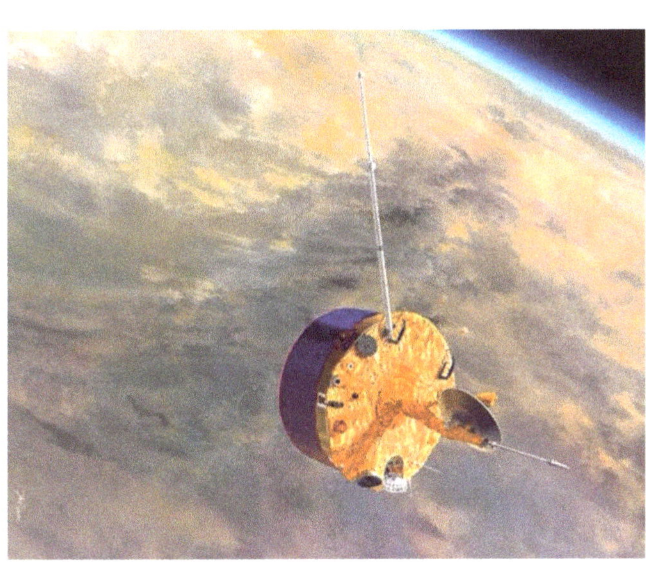

Pioneer Venus Orbiter

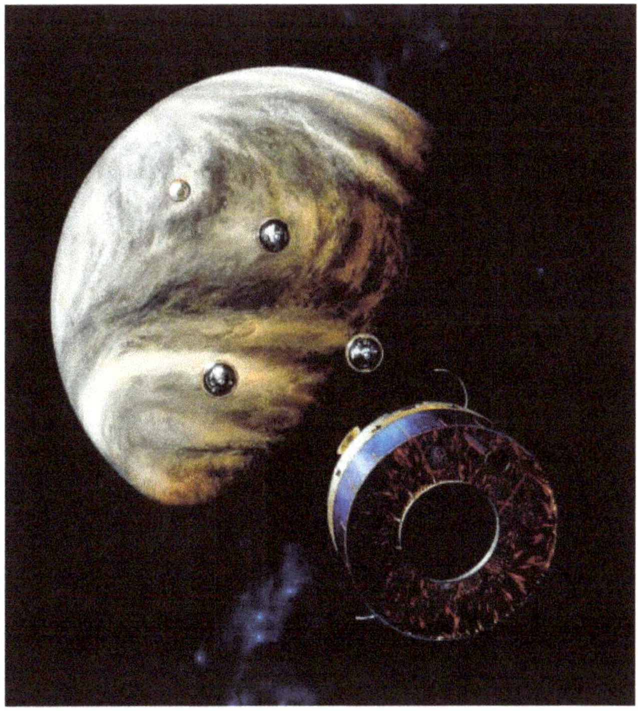

Pioneer Venus Multiple

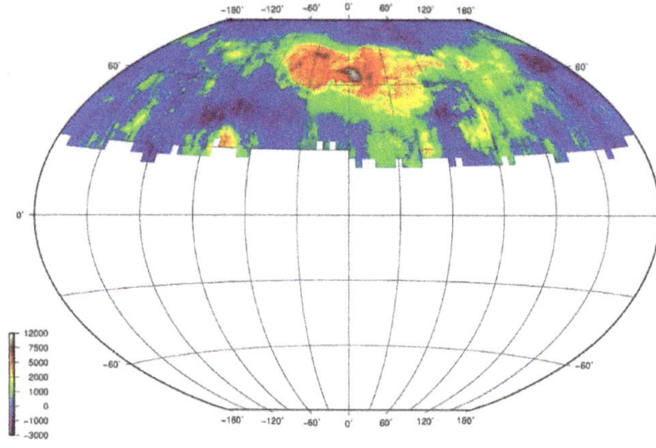

Venera 16 map of northern Venus

became part of the "Halley Armada," along with a probe from the European Space Agency and two from Japan. Each Vega had two parts, one to land on Venus, the other to fly by Halley. The craft were powered by pairs of large solar panels. On 1985 June 9 and 13, the Venus parts separated. Each included a lander much like the six previous Veneras, and a balloon. They reached Venus on June 11 and 15. At about 60 km above the surface, a parachute opened, and each balloon, yanked out of its compartment, inflated and went back up a bit by dropping its ballast. In the most active middle layer of the atmosphere, the balloons drifted west with the 1,250 km/hour wind, from the night side into the day and 40 percent of the way around Venus before communication was lost. The Vega 1 lander came down in Mermaid Plain, on the night side of the slow-rotating planet, so no images could be taken. There were many experiments, though some were triggered prematurely in turbulent atmosphere just before arrival. Each main bus, or flyby vehicle, had its course bent by Venus's gravity so as to intercept Halley. By 1986 March 4 and 7, they were able to take photos of it from distances of millions of kilometers. They passed nearest to it, 8,889 and 8,030 km, on March 6 and 9. Particles from the cloud around the comet peppered the craft but did not disable any instruments. The photos taken helped pinpoint the comet's position for the closer flyby to be made by the Giotto mission on 1986 Mar. 14. After the Halley encounter, the Vegas continued in orbits that roughly touched those of Venus and Earth. Vega 2 was to have flown by asteroid 2101 Adonis, but did not have enough fuel for that course correction. Instead the Vegas flew through the dust trails in the orbits of comets 72P Denning-Fujikawa, 289P Blanpain, and the disintegrated Biela's Comet.

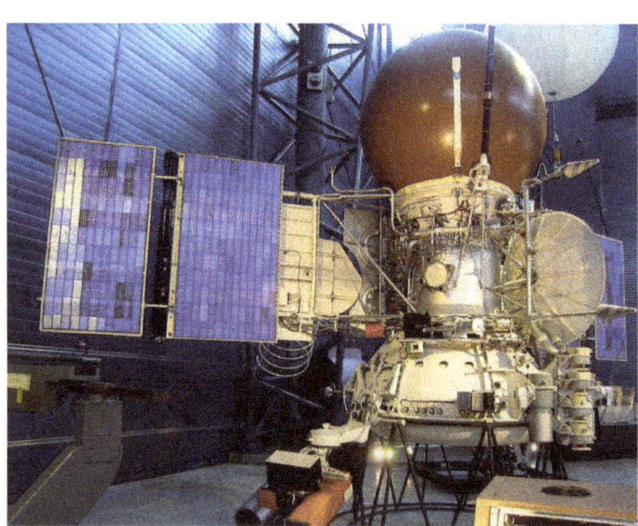

Vega 1

1989 May 4; NASA; **Magellan**, or Venus Radar Mapper. This ended an eleven-year gap in U.S. interplanetary missions. Scientists had pushed for a Venus Orbiting Imaging Radar (VOIR), which in 1982 was turned down as too expensive. In 1983 a modified version was accepted, and in 1985 was renamed for the Portuguese admiral who first circumnavigated Earth, mapping coastlines. Interplanetary ship Magellan was the first to be launched from the Space Shuttle. It reached Venus on 1990 Aug. 10, and went into a three-hour elliptical orbit outside the atmosphere, with lowest and highest points (periapsis and apoapsis) 295 and 7,762 km above the surface. The craft had to be continually reoriented so that it could face down for 37 minutes of radar soundings while nearest, then face home to transmit the data. There was complex planning to allow for Venus's long slow day, and to avoid "overly-redundant data" for the parts of Venus covered more often. Because the periapsis was 10° north of the equator, the south pole was difficult to map at all. Beginning on 1990 Sep. 15, there were 6 mapping cycles, typically lasting about 8 months. Cycles 2 and 3 filled in gaps, till the mapping covered 98% of the surface, 21% of it in stereo relief–still the best map of Venus. In cycle 4, Magellan kept its antenna pointed to Earth, where its exact speed was measured; this revealed the gravitational field–slight increases of speed were over lumps of higher gravitation. For cycle 5, Magellan gave a first test to a long-proposed technique called aerobraking, to circularize orbits. It was lowered into the thin outermost atmosphere, which over two months slowed it. In cycle 6, the lower height and almost circular orbit allowed higher-resolution gravimetry and other measurements. The finale, in late 1994 August, was the "windmill experi-

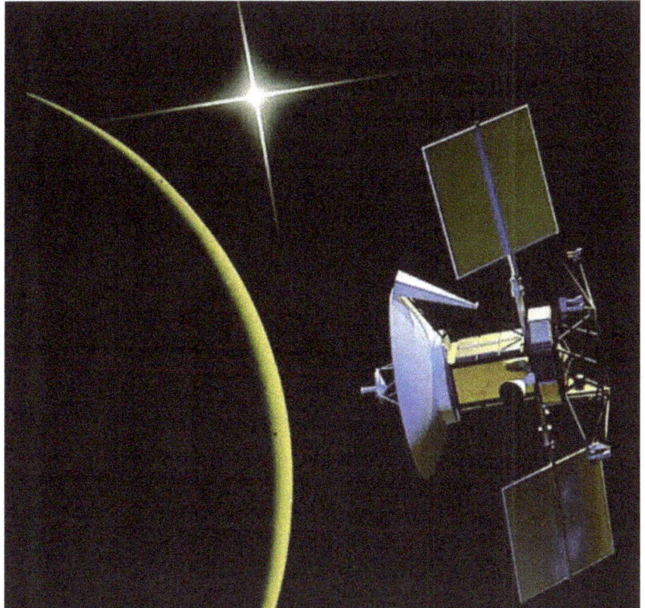

Magellan

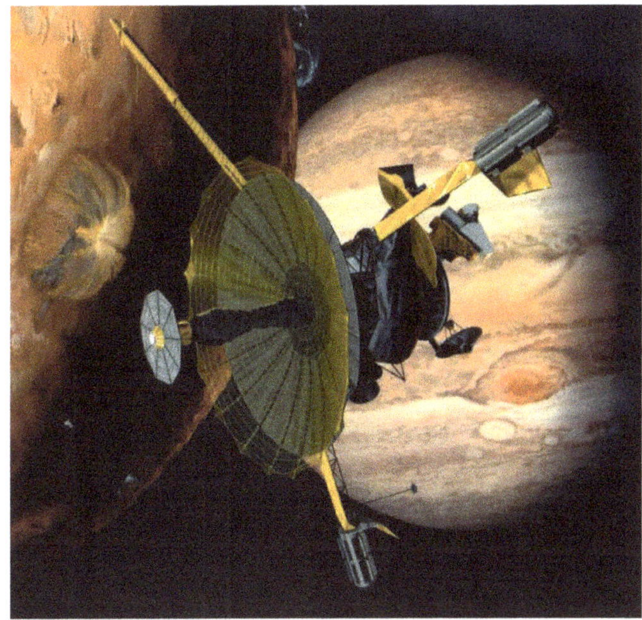

Galileo

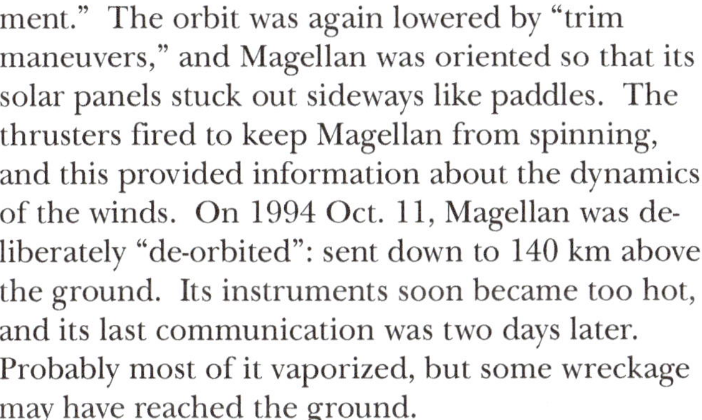

ment." The orbit was again lowered by "trim maneuvers," and Magellan was oriented so that its solar panels stuck out sideways like paddles. The thrusters fired to keep Magellan from spinning, and this provided information about the dynamics of the winds. On 1994 Oct. 11, Magellan was deliberately "de-orbited": sent down to 140 km above the ground. Its instruments soon became too hot, and its last communication was two days later. Probably most of it vaporized, but some wreckage may have reached the ground.

1989 Oct. 18; NASA; **Galileo**. Sent to study Jupiter, flew by Venus on the way. There had been more distant flybys of Jupiter by Pioneers 10 and 11 and Voyagers 1 and 2 (1973 Dec. 3, 1974 Dec. 3, 1979 March 5 and July 9), Galileo, originally named the Jupiter Orbiter and Probe, could have set out as early as 1980, using the convenient alignment of the outer planets to reach Jupiter in 2½ years, but was delayed by politics and then by setbacks of the Space Shuttle. Carried up to Earth orbit by the shuttle Atlantis. it had to take an elaborate 6-year route, pumping itself up by traveling around the Sun in three widening loops, on each of which it whipped past a planet–Venus once and Earth twice. Because the first loop was inward, protection had to be added against the heat of the Sun. The passages of Earth were on the sunward side to speed and enlarge the orbit, bending it outward; instruments studied the far side of the Moon.

Passing 16,000 km from Venus on 1990 Feb. 10, they studied the greenhouse effect and destruction of ozone.

The second and third loops were designed so that Galileo made the first flybys of asteroids: 951 Gaspra (1991 Oct. 29) and 243 Ida (1993 Aug. 28), making at Ida the first discovery of an asteroid satellite, which was named Dactyl. While Galileo was on the way to Jupiter, in 1994, it was able to make the most detailed observations of Comet Shoemaker-Levy 9's break-up and crash into the planet. Galileo reached Jupiter on 1995 Dec. 7 and became the first to go into orbit around it. Despite hardware problems, many observations were made of the Jovian atmosphere and the four "Galilean" satellites (discovered by Galileo Galilei in 1609-1610). Volcanoes were detected on Io, and oceans under the surfaces of the other three, Europe, Ganymede, and Callisto. To make sure no Earth microbes could contaminate the satellites, which just might have their own biota, Galileo was on 2003 Sep. 21 sent down at 48 km per second to vaporize in Jupiter's atmosphere.

1997 Oct. 15; NASA; **Cassini-Huygens**, or Cassini for short, consisted of the Cassini orbiter and the Huygens descender, and flew by Venus–twice–on its elaborate journey to Saturn. It was a collaboration of NASA, the European Space Agency), and the Italian Space Agency, and took its name from two of the great scientists from the seventeenth-century florescence of European astronomy. Christiaan Huygens, 1629-1695, with his improved telescope was first to discover a Saturnian satellite and resolve the nature of Saturn's rings; he invented much else, including pendulum clocks and the wave theory of light. Giovanni Domenico Cassini, 1675-1712, discovered four of

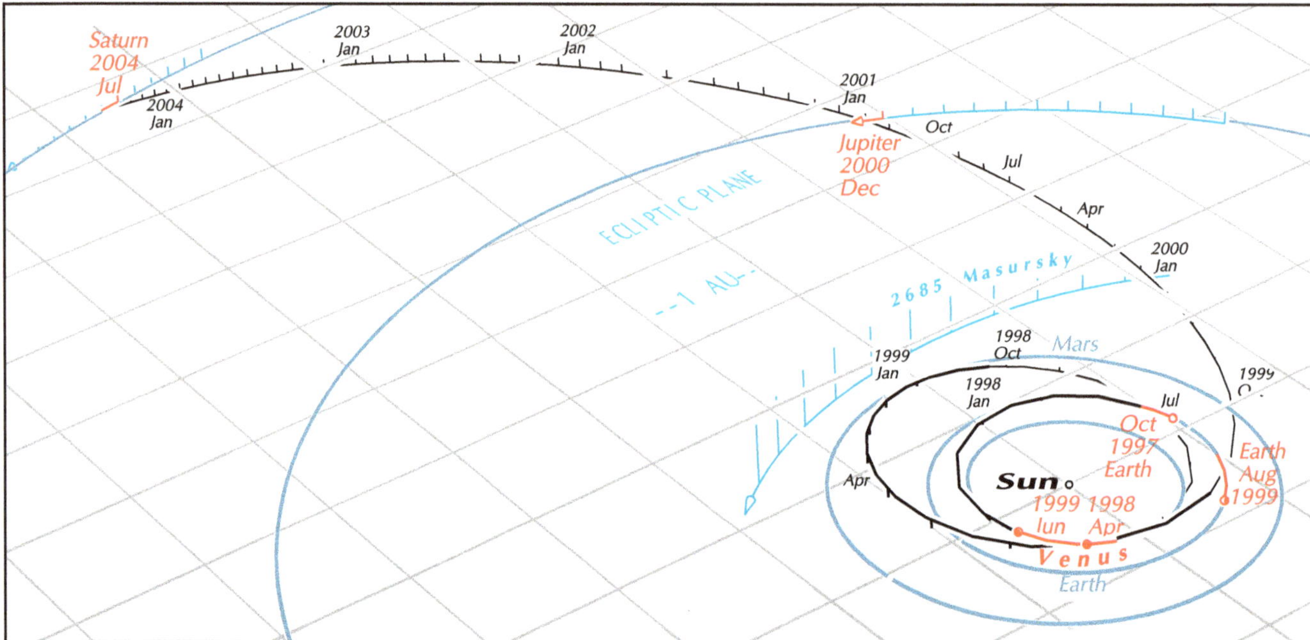

The complex journey of Cassini-Huygens. It carried 72 pounds of plutonium oxide for generating electricity from radioactive decay; and critics argued that an accident could release plutonium into the atmosphere and kill half a million people. So, with legal action pending, the Jet Propulsion Laboratory in Pasadena would not give me the information I needed to plot this diagram for my *Astronomical Calendar 1998*, and I left a blank space instead. I was later able to get it—not the orbital elements I asked for, but only positions at 20-day intervals; hence the apparent straight lines in the path.

The path was designed to take advantage of an arrangement of the planets that would not recur for 175 years. Cassini got gravitational fillips by curving first inward past Venus (1998 Apr. 26), which sent it out as far as the orbit of Mars and in past Venus again (1999 June 24, less closely); which sent it out past Earth (1999 Aug. 18, at an altitude of 725 miles); which sent it out past Jupiter (2000 Dec. 30). It used its engine to make a few course-corrections.

The picture is from a viewpoint 30 astronomical units from the Sun and 35° north of the ecliptic plane. Stalks connect the planets to the ecliptic plane at the beginning of each month. Red emphasizes the months of the Cassini encounters.

Saturn's satellites and the "Cassini" division in its rings; his son, grandson, and great-grandson were astronomers too, all like him directors of, and some born at, the Paris Observatory.

Cassini-Huygens was, to quote Clifford Cunningham's space exploration article in my Astronomical Calendar 1999: "At $3.3 billion, and 12,600 pounds, it is the most expensive and heaviest planetary mission ever." Slight glitches (including an accidental blast from an air-conditioner) caused postponement of earlier launch dates.

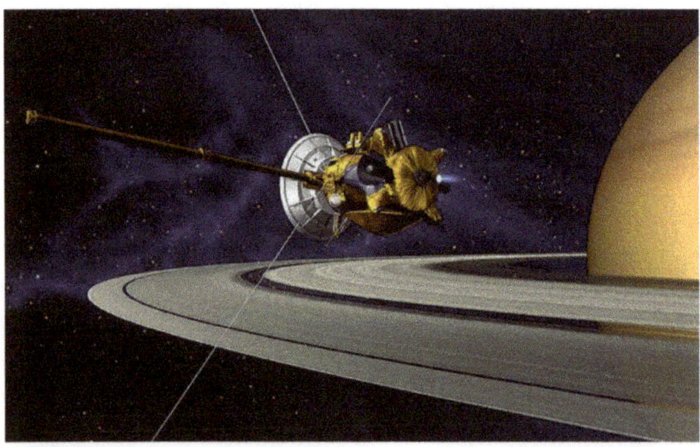
Cassini-Huygens at Saturn

It shaped the long course to Saturn by flying by Venus in Apr. 1998 and July 1999, Earth in Aug. 1999, and Jupiter in Dec. 2000, using gravitational boosts from the planets. While in the inner solar system, it did some measuring of the dust environment; and after crossing Mars's orbit it passed on 2000 Jan. 23 within 16 million kilometers (about 4 Moon-distances) of a main-belt asteroid, 2685 Masursky.

It became, on 2004 July 1, the fourth spacecraft to visit Saturn, and the first to go into orbit around it. The Huygens module separated on 2004 Dec. 25, and on 2005 Jan. 14 landed by parachute on the huge satellite Titan (larger than Mercury and with an early-Earth-like atmosphere); it relayed data by way of the orbiter for 90 minutes. This was the first landing on any natural satellite other than Earth's Moon, and the first on a solid body in the outer Solar System. The intention had been a four-year stay, to May 2008, but it was twice extended, and Cassini spent 13 years orbiting Saturn, sending rich information, and concluding with a "grand finale": a series of risky passes through the narrow gaps in the inner rings (not the largest and first-discovered gap, the Cassini Division), On 2017 Sep. 15, Cassini was sent down to burn up in Saturn's atmosphere, so as not to infect any satellite with Earth cooties.

2004 Aug. 3; NASA; **MESSENGER**. The name is both a tortuous acronym–MErcury Surface, Space ENvironment, GEochemistry, and Ranging–and an allusion to Mercury, messenger of the gods. In order to slow to its arrival at the innermost planet with least use of fuel, MESSENGER took a complex course, flying by Earth once, Venus twice, and Mercury thrice–aiming past them in such as way as to be decelerated rather than speeded. At the first flyby of Venus, 2006 Oct. 24, the planet was almost exactly at superior conjunction (Oct. 27, 17 UT); Earth was on the opposite side of the Sun. So there was no communication for about two months around that time, and no scientific observations. After another maneuver, MESSENGER re-passed Venus on 2007 June 5, at a lower altitude (338 km). This caused the greatest reduction of speed, and allowed a lot of observing of the upper atmosphere. MESSENGER first flew by Mercury in Jan, 2008–the second to do so after Mariner 10–and got into orbit around it from 2011 Mar. 18 to 2015 Apr. 30, when it was made to crash on the surface.

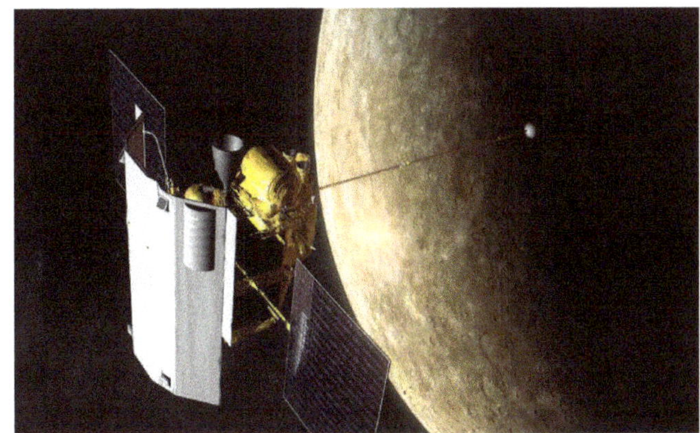

MESSENGER

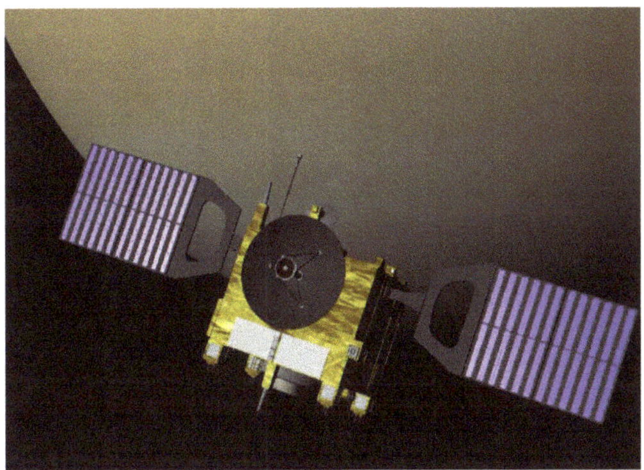

Venus Express

2005 Nov. 9; ESA; **Venus Express**, or VEX, the European agency's first Venus mission, developed from the earlier Mars Orbiter. It was launched by a Soyuz rocket from the Russians' Baikonur Cosmodrome in Kazakhstan. It arrived at Venus on 2006 Apr. 11; fired its engine to slow into a 9-day orbit and eventually on May 7 a 24-hour polar orbit, ranging from 66,000 km above Venus's south pole to 250 over the north. It made detailed studies of the surface of Venus and especially the atmosphere, helping the understanding of climate change on Earth. The mission was to last 500 days, about two of Venus's days, but was extended five times, to 2014 Dec. 16. It was last heard from on 2015 Jan. 18, and it probably crashed into Venus about the end of that month.

2010 May 20; Japan; **Akatsuki**, "dawn," or the Venus Climate Orbiter, or Planet-C. After a two-day weather delay, JAXA, the Japanese space agency, used one rocket to launch Akatsuki, the IKAROS solar sailer, the student-built Shin'en, and three other small student spacecraft, called Waseda-SAT2, K-Sat, and Negai *. Akatsuki carried about 90 aluminum plates engraved with the names and messages of 260,214 people, also a specimen of modern Japanese culture: pictures of a "super-deformed" artificial nymph with a "Vocaloid" synthetic voice, Akatsuki arrived at Venus on 2010 Dec. 6, but failed to get into orbit and flew past. Engineers had to let it circulate around the Sun for exactly 5 years, till it was in position for another and successful try, on 2015 Dec. 7. One of the first discoveries, that month, was a possible gravity wave (a reaction at the surface between two fluids). Akatsuki continuing to study Venus's cloud layers, using five cameras for different wavelengths, though two of them have failed. The main mission was completed in 2018,

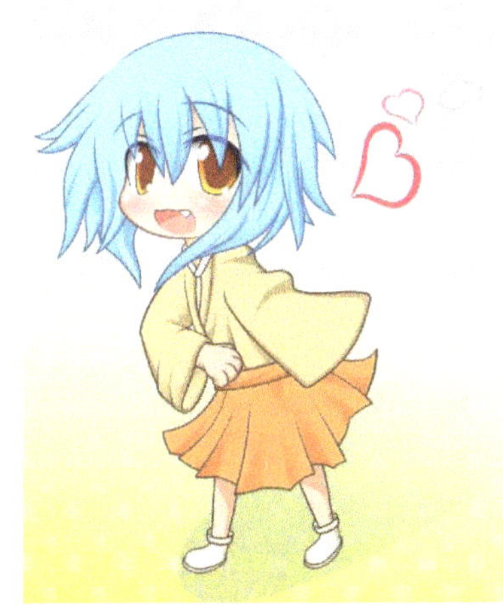

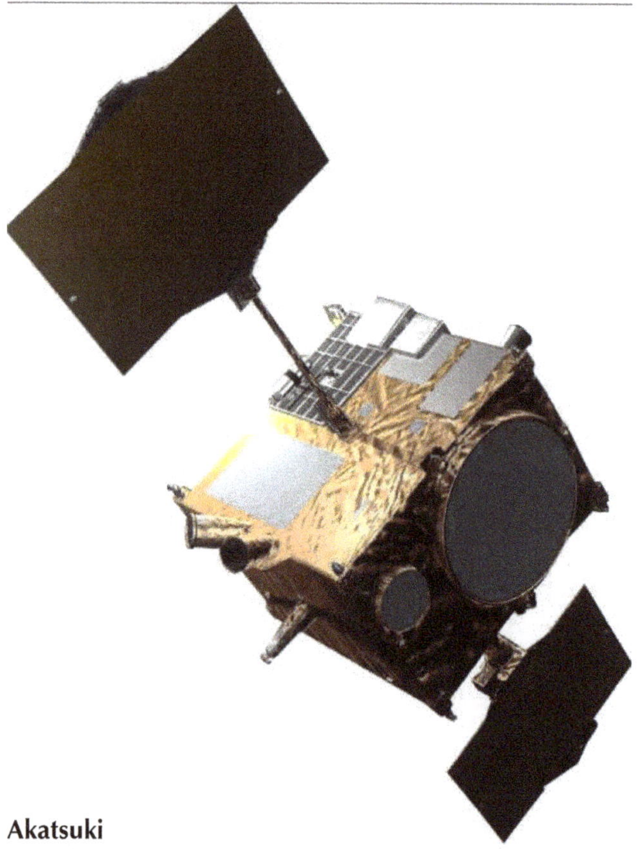

Akatsuki

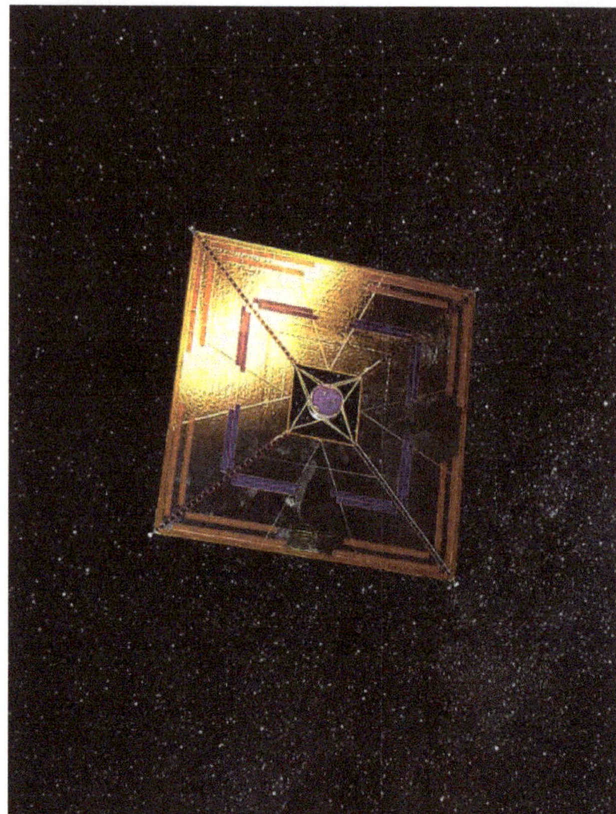

IKAROS

was extended, and may be extended again, since fuel should last till at least 2021.

2010 May 20; Japan; **IKAROS**, another effortful acronym: "Interplanetary Kite-craft Accelerated by Radiation Of the Sun." Icarus, more often known by that Latin form of his name, was the rash youth who, fitted with wings by his father Daedalus (Daidalos) for their escape from the Cretan maze, flew too near to the Sun, so that the wax melted and he fell to his death in the Icarian Sea. Launched along with Akatsuki, IKAROS is the first successful demonstration of the beautiful solar sail technology: gentle but cumulative acceleration by sunlight. The sail contains solar cells for generating power. And along its edges are 80 liquid crystal panels: by varying their reflectance, the sail's attitude can be changed without use of fuel. Other than the thin sail, the craft consists of little but two cameras, named DCAM1 and DCAM2, making it, according to Guinness World Records, the smallest inter-planetary spacecraft. It flew by Venus, at a height of about 80,800 km, on 2010 Dec. 8, two days after Akatsuki's unintended flyby, but its mission did not include observation of the planet. The extended mission continues, the craft spinning about twice a minute and making observations of interplanetary space, and the scientists refining their skills of solar sail navigation, though since 2012 the attitude control and hence the transmitting of data have been intermittent. IKAROS will be followed by a solar sail probe to Jupiter and the Trojan asteroids, in the late 2020s.

2010 May 20; Japan; **Shin'en**. UNISEC was a coalition of several Japanese universities. Students built the craft, calling it UNISEC Technology Experiment Carrier 1, or NITEC-1. When launched by the same rocket as Akatsuki and IKAROS, it took its Japanese name, meaning "profound." It was to have tested the effect of space conditions on six university-built computers, and to travel on past Venus. It did separate from the rocket into a heliocentric orbit, the first student-built craft to get that far. But contact with it was lost a few minutes later. It probably did fly by Venus, silently, in December 2010.

2018 Oct, 20; ESA and Japan; **BepiColombo**, two craft launched together: Mercury Planetary Orbiter (MPO) and Mercury Magnetospheric Orbiter (MMO). There was a flyby of Earth on 2020 Apr. 10, and there will be flybys of Venus in 2020 and 2021, and six of Mercury between 2021 and 2025, so that speed will be slow enough for insertion into orbit around Mercury on 2025 Dec. 5.

The goddess

Venus has power over love, desire, sex, and procreation. Under various names, she has been venerated in many cultures. She is sometimes, and sometimes not, identified with a supreme goddess, or a mother of all, or a primeval state of chaos before the gods. She is usually associated with, or is, the planet Venus. She is usually distinct from another goddess who stands for female power and is identified with the Moon–Artemis of the Greeks, Diana of the Romans–though the Moon deity is in some cultures male. Other spirits may preside over such aspects as lust, marriage, childbirth, motherhood, nurturing, home-making, femininity, virginity (Artemis), fertility (Ceres), witchcraft (Hecate), orgiastic excess (Bacchus). And not all erotic deities are female:

Inanna

History begins with the Sumerians of southern Mesopotamia, the land "amid the rivers," now known by the Arabic word Iraq, the "opening" or "vale." The Sumerians started perhaps as early as 3500 BC to use some form of writing. It became the elaborate script we call cuneiform, "wedge-shaped," made by jabbing a reed stylus into clay tablets.

Memory of this civilization became lost around two millennia later. In the 19th century, archaeologists revealed it and deciphered the Sumerian language, which has no known relation to any other. The name *Shumer* was applied by the neighboring Akkadian people; the Sumerians called themselves something like *ug-sag-gi-ga*, "black-headed people." Mesopotamia is dotted with *tells*, mounds built by successive layers of occupation. There were more than twenty Sumerian cities around the lower Euphrates and Tigris. The Sumerian king-lists recorded 21 dynasties that took turns to dominate, including several dynasties of Kish, Uruk, and Ur, and a dynasty "before the Flood" with reigns of fantastic length, starting with Eridu, apparently the first city. All dates are uncertain, depending on correlation with the archaeology of other regions and on observations of eclipses. Dates in the "short" chronology are 54 years later than those in the "middle" chronology.

The Sumerians broadly agreed on a large family of gods, but the cities had variant tradi-

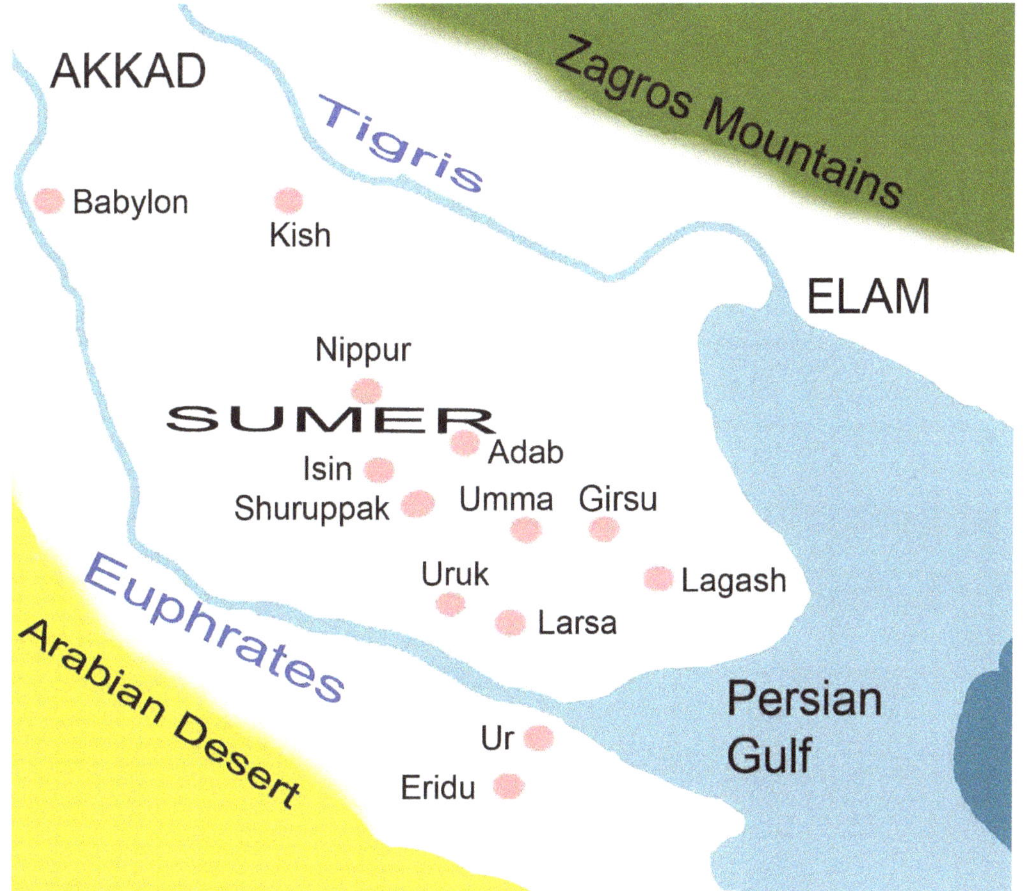

Silt from the rivers has caused the Persian Gulf coastline to recede since ancient times, as indicated approximately by the lighter blue.

tions. In their low-lying world, the first being was Nammu, the water, who gave birth to the land. Ki, and the sky, An, which comprised three concentric domes; in the lowest, made of jasper, were the stars. The ruling god was wise Enki, of the waters and of knowledge.

The love goddess was Inanna, though she was also imperious and warlike. She figures in more myths than any other of the gods. They were static; she gadded around, conquered, had adventures.

She was the brightest star, that moves around and reappears on the other side of the Sun. Her father was Nanna, the Moon, and her twin brother was Utu, the Sun. She and the Sun and Moon were the three powers of the sky. (Her name may, or may not, be related to Sumerian words nin, "lady," and an, "sky.") Her symbols were a lion, a hook-shaped bundle of reeds, and, mainly an eight-pointed star, presumably because the brightest of "stars" shows the most noticeable diffraction spikes.

She probably started as the patroness of Uruk (called Erech in the Hebrew book of *Genesis*), where was her chief temple, the E-anna. "house of heaven." Uruk, and Inanna, extended their power over Sumer and beyond during what archaeologists call the early Uruk period. Some think the name of Uruk is preserved in Iraq.

Inanna's consort, or primary consort, whom she often treated roughly, was her brother Dumuzi.

In the great myth of "Inanna's Descent into the Underworld," she intends to conquer it from her elder sister Ereshkigal, as she had usurped the domains of several other deities. But the seven judges of the Underworld condemn her for arrogance, and she is struck dead. But she had an attendant, Ninshubur, "Queen of the East," who was also a messenger between the gods, and who went up to plead for her. The gods declined to help, except Enki, who sends two spirits to lead Inanna back to the world of the living. The gatekeepers of the Underworld seize her husband Dumuzi and drag him down instead of her. It was decided that Dumuzi should spend half of the year in the Underworld and half in the Over, alternating with his sister Geshtinanna.

We may guess that Inanna's disappearance into and reappearance from the Underworld symbolized, or even was suggested by, her planet's disappearance over the western horizon and reappearance in the east some mornings later. There is some evidence that the Sumerians understood the identity of the evening and morning stars as early as about 3000 BC.

Sumerian literature is rich in variant stories. Inanna's father may be Nammu, the primeval sea; or Enki himself. From Enki she steals the me, the rules of civilization. Her husband may be Nergal, a god of the Underworld. In a poem called "Inanna Prefers the Farmer," Dumuzi and farmer Enkimdu vie to marry her and she chooses the latter. She is raped in her sleep by a gardener named Shukaletuda, whom she punishes. Dumuzi is murdered by a bandit woman called Bilulu, whom Inanna pursues and executes. Inanna and Dumuzi are associated with the constellations Pisces and Aries.

And Dumuzi appears twice in the king list: in the legendary antediluvian dynasty (where he is called "the shepherd" and reigns for 10 sar, that is, 36,000 years), and in the first dynasty of Uruk

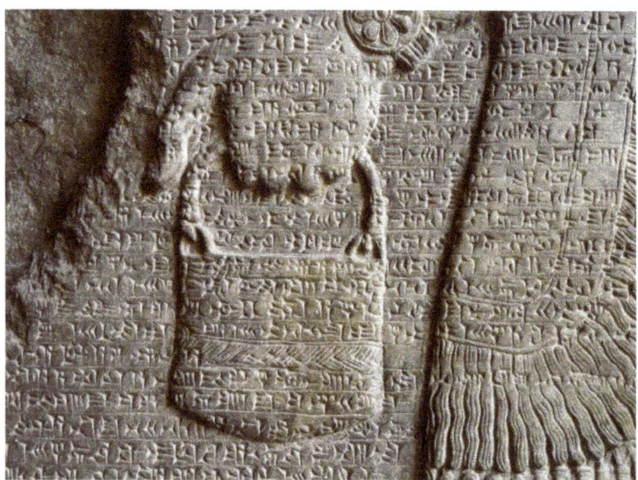

Cuneiform text covering a sculptured surface.

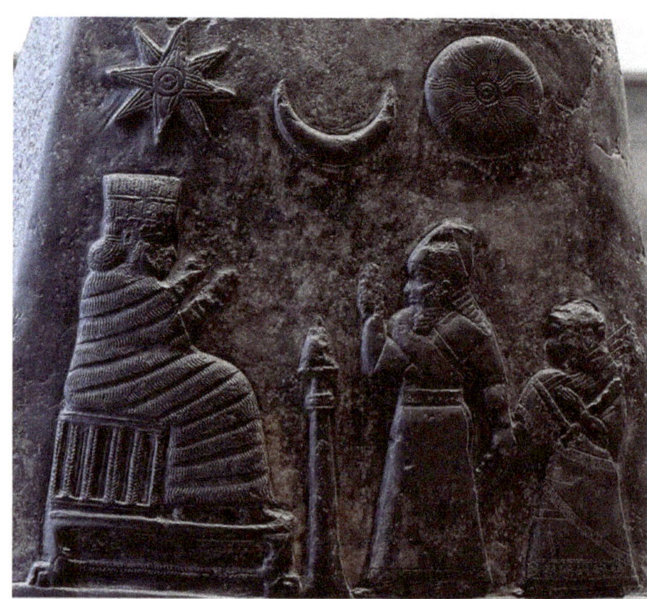

Inanna, heavily robed, and surmounted by her symbol the eight-pointed star, queens it over her brothers the Moon and Sun.

(where he is called "the fisherman" and reigns a mere century).

Later kings of Uruk may have acted the role of the deified Dumuzi at an annual spring ceremony, "marrying" the high priestess of Inanna.

The Sumerians disappeared, or rather their language did. Possibly descended from them are the Marsh Arabs of southernmost Iraq, whom Saddam Hussein tried to destroy.

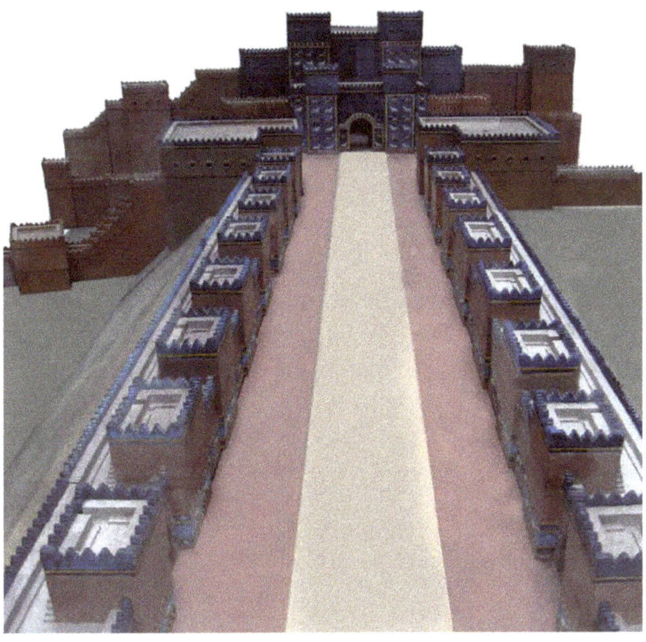

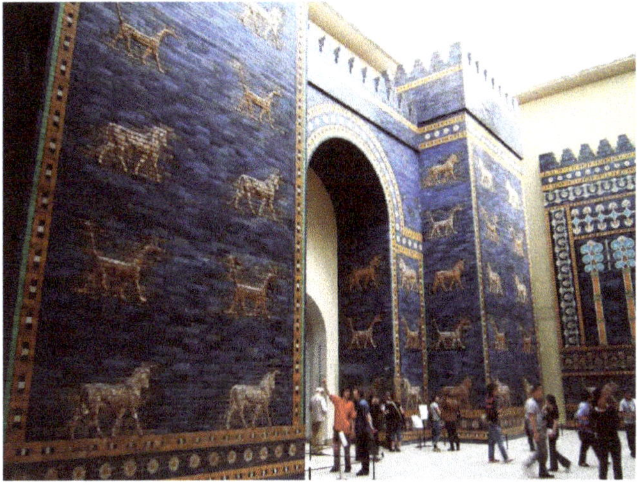

Babylon's procession Ishtar Gate, built under Nebuchadnezzar II, about 575 BC. A model; a reconstruction in Berlin's Pergamon Museum.

Ishtar

Sharing Mesopotamia with the Sumerians were people we call the Akkadians. They spoke an entirely different language, of the Semitic family that includes Hebrew, Aramaic, Syriac, Phoenician, Arabic, Maltese, and the languages of central Ethiopia.

These Semitic people lived partly mingled with but mostly to the north of the Sumerians, in a kind of symbiosis, adopting cultural traits and the cuneiform writing system. We call them Akkadian because they rose to power about 2350 BC under Sargon of Akkad (or Agade), a city in central Mesopotamia but of unknown location, probably not far north of modern Baghdad.

For the Akkadians, Utu the Sun became Shamash (the same word as Arabic shams); Nanna the Moon god became Sin; Ninshubur the messenger became a male deity, Papsukkal (perhaps like the Greeks' Hermes); Dumuzi became Tammuz. And they identified Inanna with their Ishtar.

This name does not have the Semitic feminine ending. A theory is that there were originally a love goddess of the Evening Star and a war god of the Morning Star, until it was understood that the two are the same. It may be that, when the roles were combined, the deity was of ambiguous gender, and the Sumerians settled on the feminine name and the Akkadians on the masculine. Inanna-Ishtar's temples had a staff of androgynous and cross-dressing people called galla.

Sargon sent his daughter Enheduanna to be high priestess of Inanna in Ur. She composed a hymn to the goddess she called both Inanna and Ishtar, and other poetry–the first in history by a poet whose name is known.

After the Sargonid empire, there were further times of Sumerian dominance with dynasties of Uruk, Ur, and Isin, but from about 1900 BC history was largely an alternation of empire between Babylon (fifty miles south of Baghdad), and Assyria in the north–with irruptions of tribes from the mountains on the east and north: Guti, Hurrians, Hittites, Kassites. Mesopotamia fell under the Median empire of Cyaxares in 626 BC, the Persian empire of Cyrus in 539, and the Macedonian empire of Alexander in 331.

As Ishtar, the goddess rose to the top of the pantheon. She had temples all over Mesopotamia. The Assyrians especially favored her, ranking her above their national god Ashur. By the process called syncretism, at least nine lesser or local goddesses were absorbed into her–came to be aspects of or names of Ishtar.

Ishtar appears in the popular epic of Gilgamesh. He was a king of the First Dynasty of Uruk, successor to Dumuzi "the fisherman." But the epic was composed in Akkadian, in several versions which were eventually woven together by a single scribe. Many exploits were attributed to Gilgamesh, including a descent into the netherworld. He helps Ishtar get rid of a pestilence, and she rewards him with a gift of some magic objects. She asks him to be her lover; he declines, listing the fates of predecessors in that role; she sends against him the Bull of Heaven; he slays it; the gods administer the punishment of mortal disease, but the one who dies is his comrade, the wild man Enkidu.

Astarte

The goddess became Astoreth–with the Semitic feminine ending–for the people of Syria, Canaan (Palestine), and Phoenicia (the modern Lebanon), and was carried by the sea-faring Phoenicians to the greatest of their colonies, Carthage (in what is now Tunisia).

The biblical books of Genesis and Amos mention a city in the land of Bashan (now northern Jordan and southern Syria) called Ashteroth Karnaim, apparently meaning "the Ishtars of the two horns." And in modern demonology, Ashtaroth (male) is a Duke of Hell, alongside Beelzebub and Lucifer.

The Syrian and Carthaginian goddess was referred to by the Greeks and Romans as Astarte. But this was many centuries into the historical era; the concept of the love goddess had long existed in Greece,

Ze eshq-e nâ-tamâm-e mâ jamâl-e yâr mostaghnî ast.
Be-âb o rang o khâl o khat che hâjat rû-ye zîbâ-râ?

From our imperfect love the beauty of the Friend is independent.
What need has the fair face of paint or adornment?

–Hâfez (1315-1390), one of the lines of his *ghazal* or ode that begins "If that Turk of Shiraz." The *yâr* or ideal friend praised in Persian poetry is sometimes–perhaps–an allegory for God.

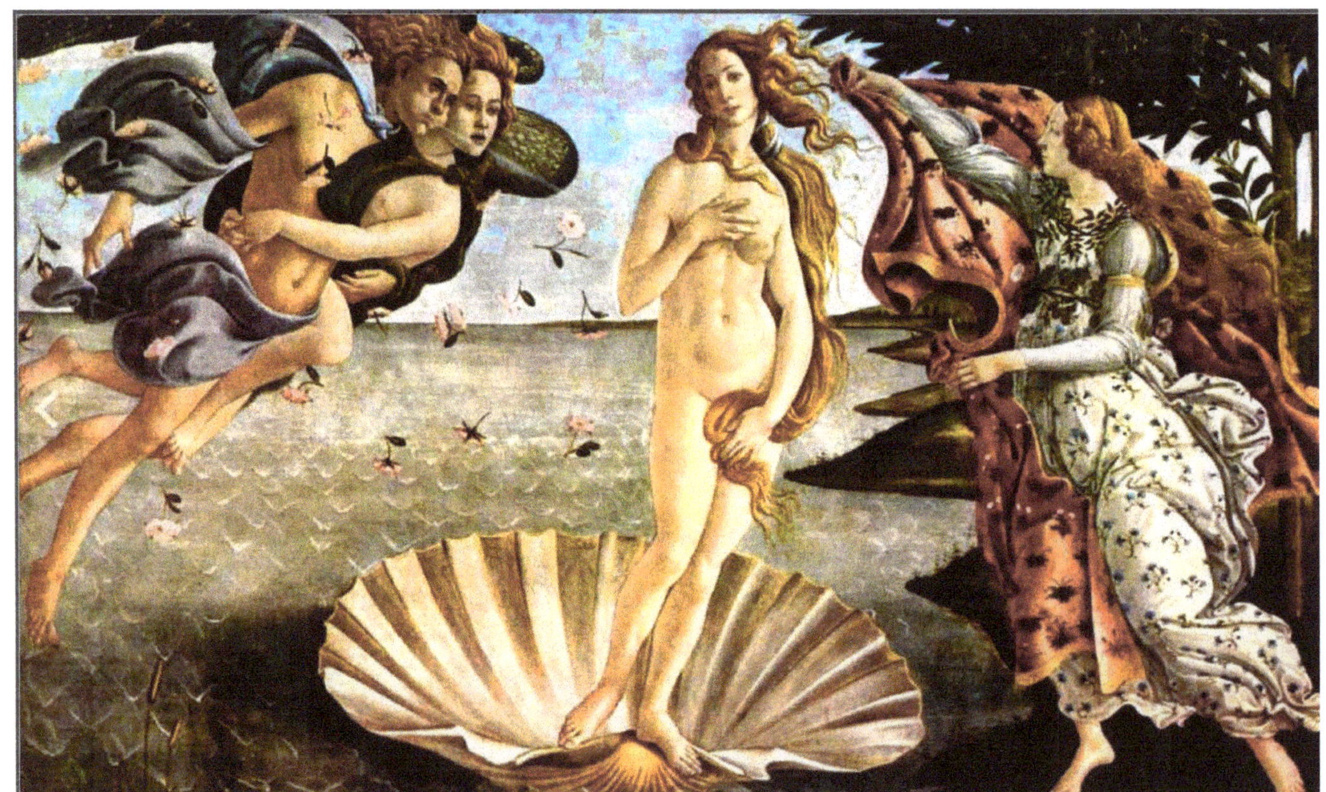

Sandro Botticelli, The Birth of Venus, circa 1485, tempera on canvas; in the Uffizi, Florence

Aphrodite

From Syria, probably, the cult of an Ishtar-like goddess spread to Cyprus, and onward along the Mediterranean sea to the Minoan civilization, which flourished on Crete from about 2700 to 1100 BC, and perhaps onward to the Mycenaean civilization of the Greek mainland. Toward the end of the Bronze Age, new migrations from the north caused Mycenaean civilization to collapse, and there was an eddying of "Sea Peoples" around the Mediterranean. Greeks colonized Cyprus around 1100 BC. The goddess became Aphrodite, meaning that she was born from the foam, aphros, of the sea. Her temple at Paphos in Cyprus remained the chief seat of her worship. Myths embodies her attributes. She was attended by flights of doves and sparrows, birds that are often seen courting and mating. Her girdle, the Zone of Aphrodite, when laid around her hips or anyone else's, made the wearer irresistible, so she was loth to lend it to her sister goddesses. Being sea-born, she came ashore riding a scallop shell. A more atavistic version of her birth was that the first chief god, Uranus of the sky, was overthrown by his son Cronus, leader of the generation of

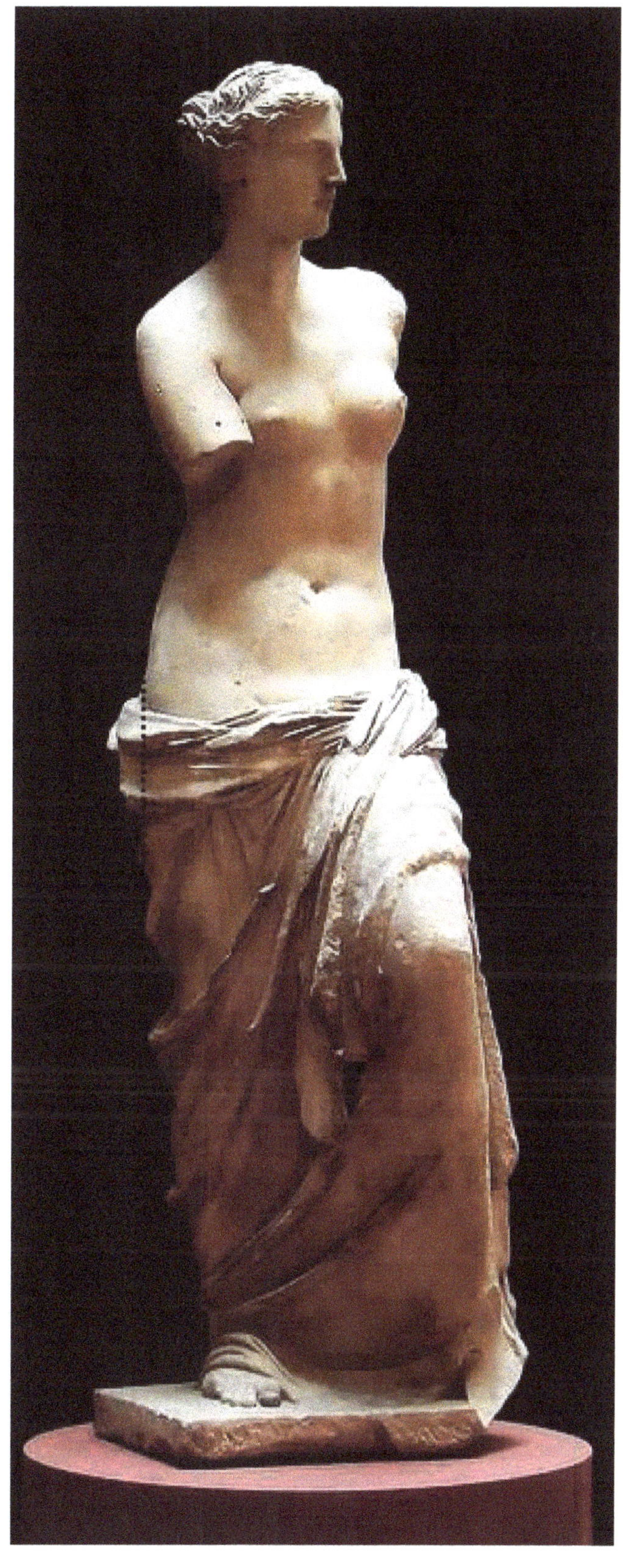

The Venus de Milo; marble, slightly taller than life size. The most famous image of Aphrodite, and long assumed to be the work of the most famous of Athenian sculptors, Praxiteles, but now believed to be by Alexander of Antioch, some time before 100 BC. Found on the island Melos in 1820; after discovery the original plinth and the arms were lost. Some think it represents not Aphrodite but Amphitrite (consort of Poseidon and queen of the sea). Now in the Louvre Museum in Paris.

Praxiteles used the courtesan Phryne as model for a sculpture of Aphrodite, commissioned by Cos, an island state off what is now southwestern Turkey. The Coans objected–nude statues had previously been male–and ordered a clothed version. Cnidus, on the nearby coast, cannily bought the original, and it made their city a tourist destination. The Aphrodite of Cnidus (now lost except for copies) was so beautiful that the goddess was jealous–or had herself been the model. Epigram 17 of the 18 dubiously attributed to Plato:

Hê Kupris tên Kuprin eni Knidô eipen idousa,
Pheu, pheu, pou gumnên eide me Praxitelês?

Said the Cyprian when she saw the Cyprian in Cnidus:
"Woe, woe! where Praxiteles saw me bare?"

Or, in a smoother and less literal translation suggested by my friend Geoffrey Jackson:

Venus exclaimed, when she her likeness viewed,
"Praxiteles, where did you see me nude?"

A later epigram, by Antipater of Sidon, had her listing mortals who had so viewed her: Paris, Anchises, Adonis–Praxiteles.

Titans, who castrated him and threw his genitals into the sea, causing the foam from which Aphrodite arose. This made her in a way the sister of the third chief god, Zeus, displacer of Cronus. Or Zeus was said to be her father, by Dione, a Titaness. Dione was the name of several others, and a title of Aphrodite herself, but was a sort of generic goddess name: *di-* is the stem of *Zeus* (genitive dios), *-onê* an ending for "queens."

Between Crete and the south-eastern tip of mainland Greece is a small island, Cythera, so Aphrodite was said to have first come ashore there, before settling in Cyprus. Hence two of her epithets, alternative ways of referring to her: Cytherea and Cypris. Zeus gave Aphrodite in marriage to Hephaestus, the lame smith god (Latin: Vulcan). Smiths were so vital to Iron Age communities that they may have been lamed to prevent them from moving away. Unlikely to be faithful, Aphrodite dallied with the macho war god Ares (Latin: Mars). Hephaestus caught Aphrodite and Ares at it, cast over them a cunning net he had made, and summoned King Zeus to witness their guilt.

The sight had a predictable effect on the male gods, and several of them sought Aphrodite's favors. To Hermes (Mercury) she bore Hermaphroditus, of uncertain or ambiguous or divided or alternating gender, and to Dionysus (Bacchus) the ithyphallic Priapus. The children she presented to her husband were really by Ares: Phobos, Deimos, and Harmonia. The names of the first two, meaning "fear" and "dread," have been given to the satellites of planet Mars. Harmonia's name is ironic: she became the wife of Cadmus, founder of Thebes, thus ancestress of many tragic figures: Pentheus, torn limb from limb by Bacchanalian revellers including his mother; Oedipus, who unknowingly married his own mother; the brothers Eteocles and Polyneices, whose feud led to the devastation of Thebes.

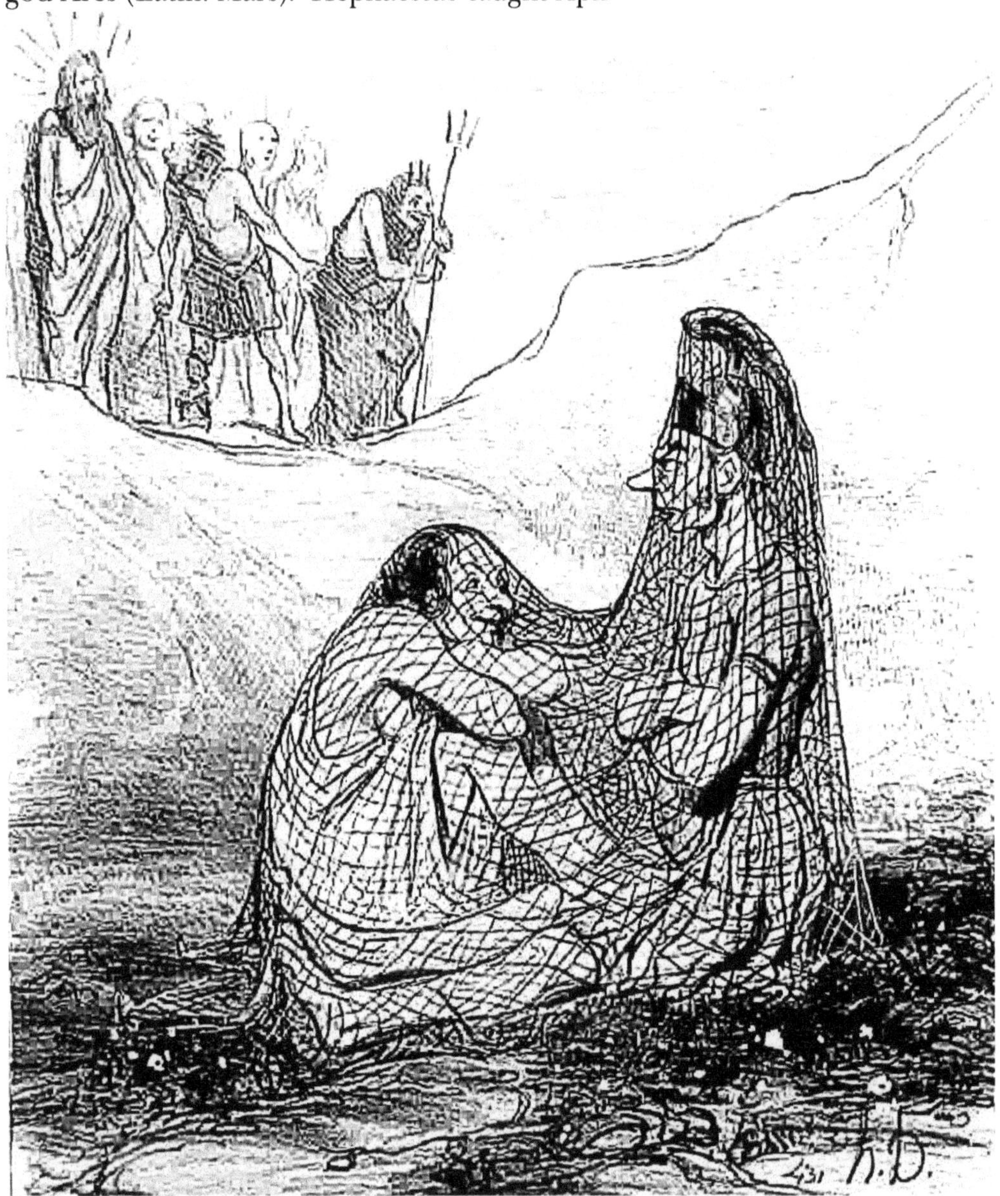

Honoré Daumier, "Mars and Venus in the Net," 1842. Jokes are not funny when repeated, right? There are two exceptions, that make me laugh every time I think of them: Lewis Carroll's parody "You are Old, Father William," and this Daumier cartoon. Vulcan indignantly displaying his evidence; Jupiter wondering how to be judicial; Neptune convulsed with glee; Mars (helmet still on) scowling; and Venus, smirking.

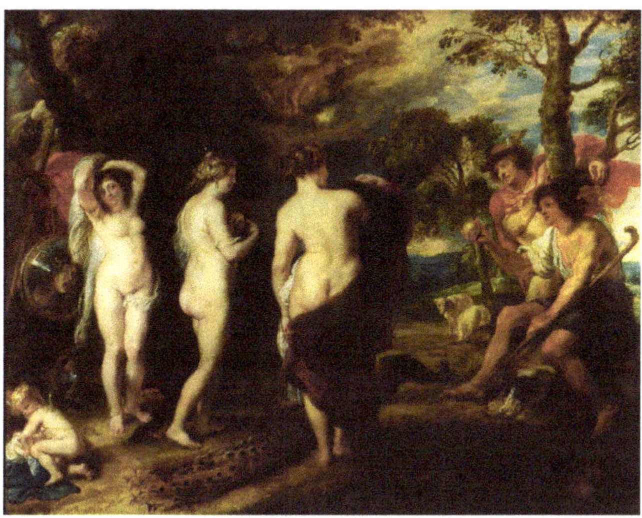

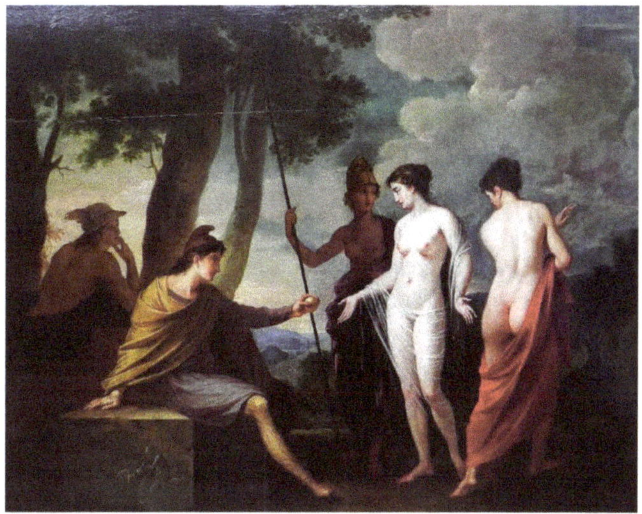

The Judgment of Paris, by Peter Paul Rubens, 1636, and less-known Joseph Hauber, circa 1818

Aphrodite among the mortals

In the Greeks' traditions about their ancestral heroic age (corresponding to the Mycenaean period of centuries earlier), there were several collective endeavors, involving many personages, and described in cycles of epics, later in poetic dramas: the hunting of the boar of Calydon; the voyage of the Argonauts to steal the Golden Fleece from Colchis (Georgia); the two sieges of Thebes; and the two sieges of Troy. Troia, also called Ilion, commanded the straits from the Aegean Sea to the Black or Euxine Sea, so the real cause of war may have been strategic. Troy was briefly seized by Heracles (Latin: Hercules) in the course of the Argo expedition. The second siege is the subject of the two completely surviving epics, the Iliad and the Odyssey, attributed to Homer. In the legendary chain of causes leading to the war, Aphrodite had a main role.

At the wedding of Peleus, a Greek chieftain, and Thetis, a sea nymph, who were to be the parents of Achilles, the gods were invited, but not Eris, spirit of strife and jealousy. She got her revenge by rolling into the feast a golden "apple of discord," inscribed *têi kallistêi,* "For the Fairest." Hera (queen goddess), Athena, and Aphrodite quarrelled over it (but not the other female members of the Olympian Twelve, Demeter, Hestia, Artemis). For a judge, they picked Paris, a shepherd in the countryside behind Troy. He was actually Alexandros, a son of Troy's king, Priam; at his birth it was foretold that he would cause Troy's ruin, so he was exposed on Mount Ida to die, but a shepherd family found and adopted him. Aphrodite won the naked beauty contest by whispering to Paris that she would help him to seduce the most beautiful of women: Helen. But she was already wife to Menelaus, king of Sparta. With Aphrodite's assistance, Paris revealed himself in Troy, sailed to Greece, became a guest at Sparta, and eloped with Helen. Menelaus's brother Agamemnon was high king of Mycenae, and married to Helen's sister Clytaemnestra. Many other princes had been suitors for Helen, and had had to vow that, if she was ever stolen from the winning suitor they would help to get her back. So an armada under Agamemnon sailed across the Aegean sea and beset Troy for ten years.

Aphrodite had another involvement with Troy. For mischievously tempting Zeus himself, she was made to fall in love with a mortal, Anchises, prince of the Dardanians (the country cousins of the Trojans) and lived humbly with him for a while in his hunting cabin. (His name begins with the preposition *ankhi*, which is like French *chez*, so he apparently "lived with Isis." The Greeks may have equated that Egyptian goddess with their Aphrodite.)

For inadvertently mentioning to drinking companions that he lived with a goddess, Anchises was struck by Zeus with a thunderbolt. Aphrodite saved his life by deflecting it, but it crippled him. Her child by him was Aineias (Latin: Aeneas), leader of the Dardanian contingent of allies at Troy.

During the 35 days of battle recounted in the *Iliad*, the gods took sides, mostly by inspiring individual warriors, but sometimes actively intervening. Aphrodite of course sympathized with Troy (as have many down the ages). The sides agreed on a duel between Paris and Menelaus; Aphrodite, seeing that Paris was getting the worst of it, wrapped him in a mist and whisked him to safety; so battle resumed. Seeing her son Aeneas attacked by Diomedes, second only to Achilles as a slaughterer, she came to the rescue, but, pricked by the Greek's spear, fled back to Olympus. Apollo, also protective of Troy because he had helped to build its walls, warned Diomedes not to fight with gods.

The Greeks at last broke into Troy by the trick of the Wooden Horse. There are many legends of the Nostoi or "returns" of Greek leaders to their homes—the surviving example is the Odyssey—and of the escapes of Trojans. Cities, tribes, and whole nations wished to believe that they were descended from one of these heroes (the British from a Trojan called Brut). One of the refugees from the burning city was the ever "pious" Aeneas, with his crippled father Anchises on his shoulders. He sailed to Carthage, refused the love of its queen, Dido, because his destiny was to become ancestor of the founders of Rome.

This legend of the descent of the Romans—or at least their patrician families such as the Julii—from the goddess, and their predetermined feud with Carthage, was narrated in the Latin answer to the Odyssey: the Aeneid, which Virgil began in 29 BC and took ten years to complete. Among celebrations of it in later European art is Purcell's opera *Dido and Aeneas.*

Was this the face that launch'd a thousand ships
And burnt the topless towers of Ilium?
Sweet Helen, make me immortal with a kiss!

—Christopher Marlowe, *The Tragical History of the Life and Death of Doctor Faustus.* Faust, enabled by Mephistophilis, has conjured up Helen's ghost. The play had such an effect that audiences believed real devils appeared on stage. The performance was close to the time of 29-year-old Kit Marlowe's mysterious murder in 1593, which made way for his contemporary Shakespeare as leading dramatist of the time. Textbook writers trying to interest students in the metric system have fun with the "thousand ships": a "millihelen" is a unit of beauty sufficient to launch one ship. (Or spaceship?)

According to another Roman legends, Diomedes, that formidable Greek who had nearly killed Aeneas, sailed away seeking a life of peace, to the Lucanian region between the toe and heel of Italy. He built a temple to Aphrodite, hoping to placate her for the woe he had caused to her beloved Troy. His colony, Aphrodisia, was taken over by the Romans in 29 BC, and called by them Venuia; it is now Venosia.

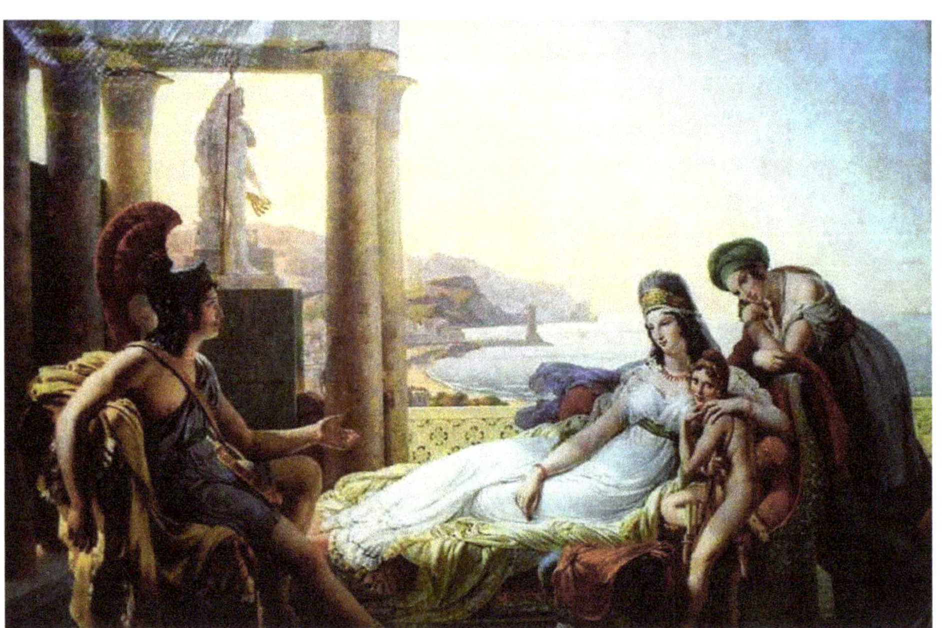

Aeneas Recounting to Dido the Disasters of Troy, by Pierre-Naarcisee Guérin (1774-1833)

Fear no danger to ensue,
The hero loves as well as you!

—Aria from Henry Purcell's opera Dido and Aeneas. Dido's maids try to reassure her. I first heard this wonderfully craggy melody in wonderfully craggy German:

Heg nich Furcht, dir stösst nichts zu,
Dein Held liebt auch ganz so wie du!

Didon dîna, dit-on, d'un dos d'un dodu dindon.

—French tongue-twister, or, I would say, tongue-slitherer, since it slides off the tongue like a peal of bells. "Dido dined, they say, on the back of a plump turkey."

Eros

Eros, the Romans' Cupid, is a male erotic godling. The name (pronounced in Greek with short first vowel and long second, in British English the other way around) is a word for sexual passion; the Latin name, Cupido (long second syllable), comes from *cupere*, "to covet."

Eros was, in the *Theogony* ("Origin of the Gods") of the poet Hesiod, around 700 BC, a primeval principle, along with Chaos and Earth. Without Eros the first gods could not have been born. In later more sentimental literature, he is a child of Aphrodite, a mischievous putto who flies about on golden wings, shooting barbed arrows; whoever is pricked lusts for whatever he or she sees. Child abuse, in modern terms.

Adonis

Imported from the east like Aphrodite, her mortal lover Adonis became for Greek women a male erotic symbol.

His name is surely the Semitic word *adôn*, "lord," as in the Hebrew *Adonai*, "my lord" (a way of referring to God without uttering the ineffable name *Yahweh*. Aphrodite and Adonis correspond to the Inanna and Dumuzi or Ishtar and Tammuz of Mesopotamian myth.

In a fragment of the poetry of Sappho (about 600 BC), girls ask how they should lament the death of Adonis, and Aphrodite tells them to beat their breasts and tear their garments. That is the first known reference to him in Greek. His story was recorded in most elaboration in Book X of the *Metamorphoses* by the Roman poet Ovid (8 AD). Myrrha, wife of king Cinyras of Cyprus, boasted that their daughter Smyrna was more beautiful than Aphrodite; who punished her by making Smyrna fall in love with her own father. When Cinyras discovered who had been in his dark bedroom, he chased Smyrna with uplifted sword, but Aphrodite changed her into a myrrh tree; split by the sword, it gave birth to Adonis. Aphrodite found the baby and entrusted him to Persephone, goddess of the Underworld. He grew up so beautiful that Aphrodite and Persephone vied for him; Zeus decreed that he spend a third of each year with each of them and a third with anyone he liked. He chose to spend that third with Aphrodite also. Ares, furious that Aphrodite preferred an effeminate mortal to himself, took the form of a boar and killed Adonis in the hunt. For centuries, women passionately mourned Adonis in an annual rite, the Adonia.

Venus and Adonis (1593), the long poem in which the seductress strives to delay her lover from going out into the fatal hunt, was in Shakespeare's lifetime more popular than any of the plays.

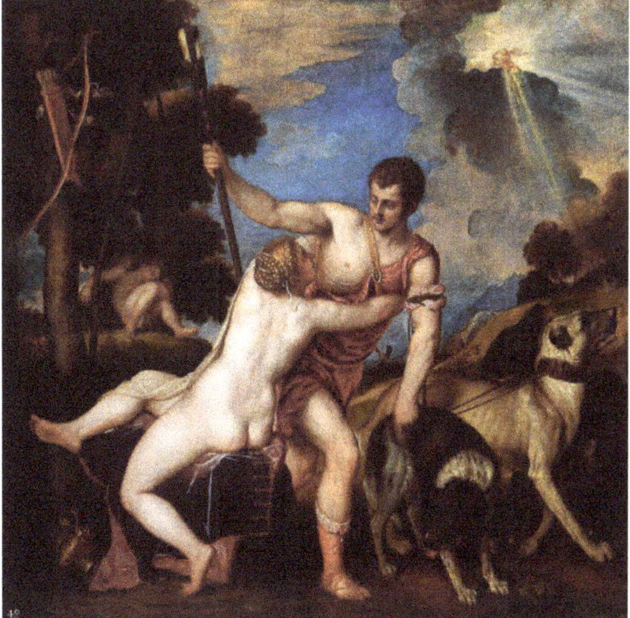

Titian, Venus and Adonis, 1554

Now is she in the very lists of love,
Her champion mounted for the hot encounter

—Beginning of the 100th stanza of Shakespeare's "Venus and Adonis." Her "annoy" will be "worse than Tantalus'" if he tears himself away to go hunting.

The name of Venus

The Romans' name for the female of females ends with *-us*, so familiar as the mark of masculine words—an amusing paradox! Moreover, it rhymes, in American though not quite in British English, with our Latinate word for the male sexual member.

But *Venus* belongs to a small group of similar nouns: *genus*, genitive *generis*, "race"; *tempus*, *temporis*, "time"; *opus*, *operis*, "work"; *corpus*, *corporis*, "body." There are many instances in Latin where an original *s* between vowels became *r*. Those other nouns are all neuter, as their plural forms show: *genera, tempora, opera, corpora.* Venus does not show her gender unless referred to in the plural, as sometimes happens (Robert Burton in *The Anatomy of Melancholy*, 1621, refers to "the graces, Veneres, pleasures, elegances").

Latin writers sometimes used the word *venus* as a mere metaphor for sex. Its deeper origin is uncertain. There may have been an Indo-European root *wen-* meaning "strive for, wish for." There may even be connections with *venia*, "forgiveness, indulgence, pardon," as in "venial sin"; or with *venire*, "to come"; or *venenum*, "venom" but also "charm, magic potion."

But not with *venari*, "to hunt." Nor with *vena*, "vein," as in *intravenous.* (Let there be no jokes about intra-Venus insertion.) Nor with English *wench.* The English words *venerate*, *venerable,* and *venereal* derive from Venus. But *venery* means hunting, not lechery.

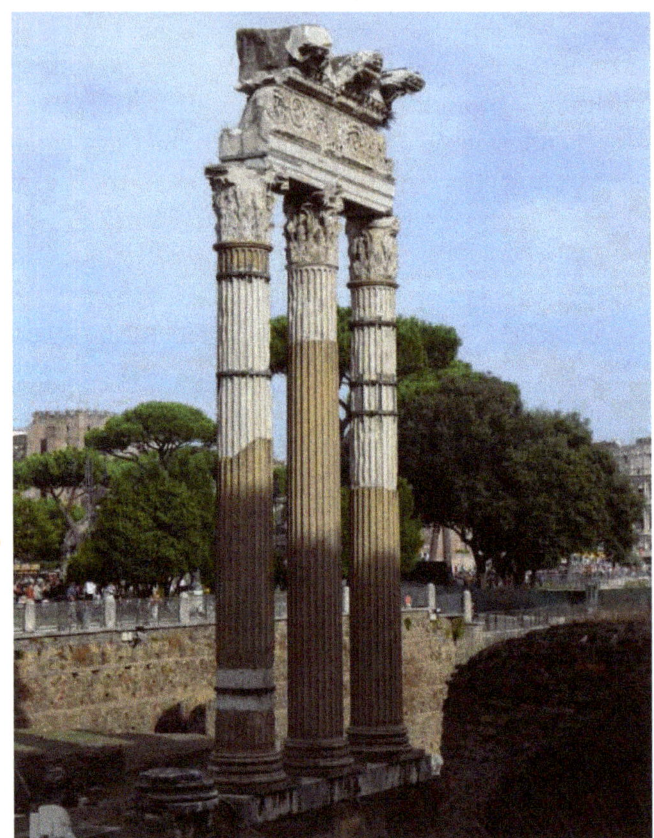

Remains of the temple of Venus Genetrix, in Julius Caesar's Forum, Rome

Salue, o venusta Sirmio!

Greetings, lovely Sirmio!

—Catullus, *Carmina*, 31, line 12. The Roman poet Catullus had a house on the Sirmio peninsula in Lake Garda.

Aphrodite becomes Venus

From early in Roman history or even prehistory, Greek colonies studded southern Italy and eastern Sicily, so that the region was known to the Romans as Magna Graecia, "great Greece." The Romans, and related Italian tribes whom they gradually absorbed—Sabines, Samnites, Umbrians, Oscans, and others—had their native god, but took to identifying them with those they learned of from the more sophisticated Greeks.

Thus Zeus became Jupiter, Hera became Juno, Ares became Mavors or Mars, and so on with the rest of the Olympians (except Apollo, who kept his name, being apparently so Greek that there was no Italian equivalent) . In later European literature, it was long customary to substitute the Latin names for the Greek—Minerva and Diana, rather than Athena and Artemis, hovered over the battlefield of Troy.

Venus took the place of Aphrodite, inherited her stories, and continued to personify love, prosperity, and good luck—including in warfare. Girls on reaching puberty offered their toys to Venus, and brides made gifts to Venus before the wedding. In dice games, the best roll was called the "Venus."

Roman Venus's first known temple resulted from a vow made in the heat of a battle with the Samnites in 295 BC. Dedicated to Venus Obsequens, "the Obliging," it was paid for with fines on prostitutes.

Rome fought three "Punic" wars against its great rival Carthage, which had been founded by the Poeni or Phoenicians. During the second war, in 211 BC, the Romans suffered a heavy defeat at Lake Trasimene, but were advised by the Sibylline oracle that they would have better luck if they had on their side Venus Erycina: by which was meant the Astarte of Eryx, a city in Sicily allied to Carthage. They attacked Eryx and carried off the idol, promising her a finer temple in Rome. This was built on the Capitoline Hill and dedicated to Venus Genetrix, the ancestress of the Romans through Aeneas.

A temple of Venus Verticordia, the "Turner of Hearts," was built in 114 BC, and many temples were dedicated to her under other epithets, such as Venus Victrix, Caelestis, Felix, the "Victorious," "Celestial," "Happy."

A form of the Greek divinity that the Romans continued to admire was Aphrodite Kallipygos, "of the beautiful rump," though they transcribed her as Venus Callipyge. A curiosity of Greek grammar was that compound adjectives take masculine endings even when feminine—*Trôades helkesipeploi*, "trailing-robed Trojanesses." But the Romans could not bring themselves to write "Venus Callipygus."

Sulla believed his successful career to be due to support from Venus. When in 89 BC he conquered Pompeii from the Samnites, he re-founded it as Colonia Veneria Cornelia (from his family name). Venus Felix did not prove lucky for Pompeii, buried by the eruption of Vesuvius in 79 AD.

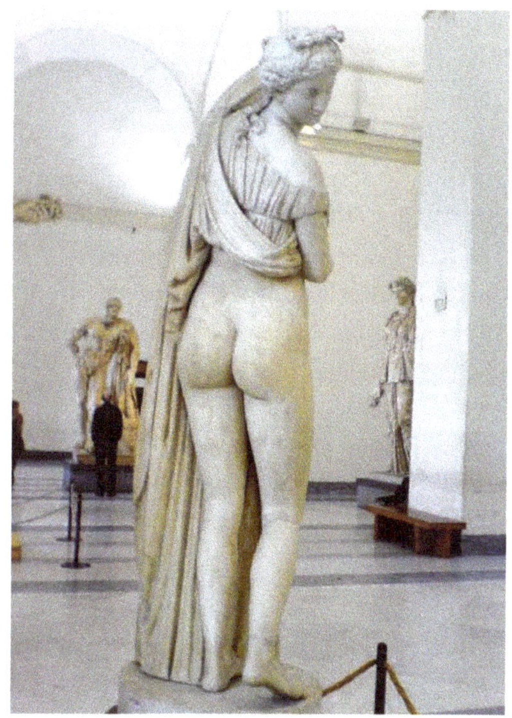

Aphrodite Kallipygos. A 1st century BC marble copy of a lost Greek original. The pose, of whichever aspect or gender, was called *anasyrma*, "up-skirt." And the goddess presenting her rear aspect was *Aversa Venus*, "Venus turned away."

Lucretius, around about 60 BC, opened his seven-thousand-line poem *De Rerum Natura*, "Of the Nature of Things," with these lines:

> Aeneadum genetrix, hominum divomque voluptas,
> alma Venus, caeli subter labentia signa . . .
>
> Progenitor of Aeneads, delight of men and gods,
> Dear Venus, who beneath the gliding signs of heaven . . .

—going on to portray her as the force that drives the winds and waves and the life in all beings. He was respecting the customs of tracing Roman ancestry to Troy and of opening epics with invocations—the *Iliad*, "Sing, goddess, the anger of Achilles"; the *Odyssey*, "Tell me, Muse, of the man of many wiles"; the *Aeneid*, "I sing of arms and the man."

However, for Lucretius, Venus was a figure of speech. His summary of physical science was based on the materialist philosophy of Epicurus, which dismissed superstitions and advised the calm seeking of happiness through not desiring too much. Venerean abandon; Epicurean restraint.

Connections

There are many parallels between the Adonis story and others. Adonis, born as punishment for a boast of beauty and found on a hillside: Andromeda, exposed for the sea monster because her mother Cassiopeia boasted of her beauty. Adonis's mother, saved by being transformed into a myrrh tree: Daphne, chased by amorous Apollo and changed into the laurel tree; Adonis dividing his year between Aphrodite and Persephone: Persephone herself, dividing her year between the Underworld and the Over, hence winter and summer. And Venus's failure to restrain Adonis from going out on the fatal hunt is reminiscent of Dido's failure to retain Aeneas in Carthage (and Calypso's failure to retain Odysseus on her island).

And many symbolisms have been discerned by the ancients or by modern scholars. Adonis may be just one of the names of a figure common to early cultures: the dying-and-rising god; or the sacred king married to the goddess and killed in alternation with his successor.

Many features of Middle Eastern mythology find echoes in the later classical world. Inanna's theft of the me: Prometheus's theft of fire for mankind. Inanna's three days of death: Christ's. Inanna's escape from the Underworld, only to lose Dumuzi into it: Orpheus, recovering Eurydice from the Underworld but at the last moment, for turning to look at her, losing her back into it. Dumuzi has to divide his year between the Over and Underworlds, as do Persephone and Adonis. The possible male and female aspects of Ishtar and the involvement of intersexual people in her cult: the hermaphrodite offspring of Hermes and Aphrodite. Ninshubur is like Iris, the rainbow, carrier of messages between Greek gods.

Others descended into the Underworld, such as Theseus and his comrade Peirithous, who had to be rescued from it by Heracles. The harrowing of Hell is a theme in many cultures. It would not be needed in a culture on planet Venus, whose surface is a hell.

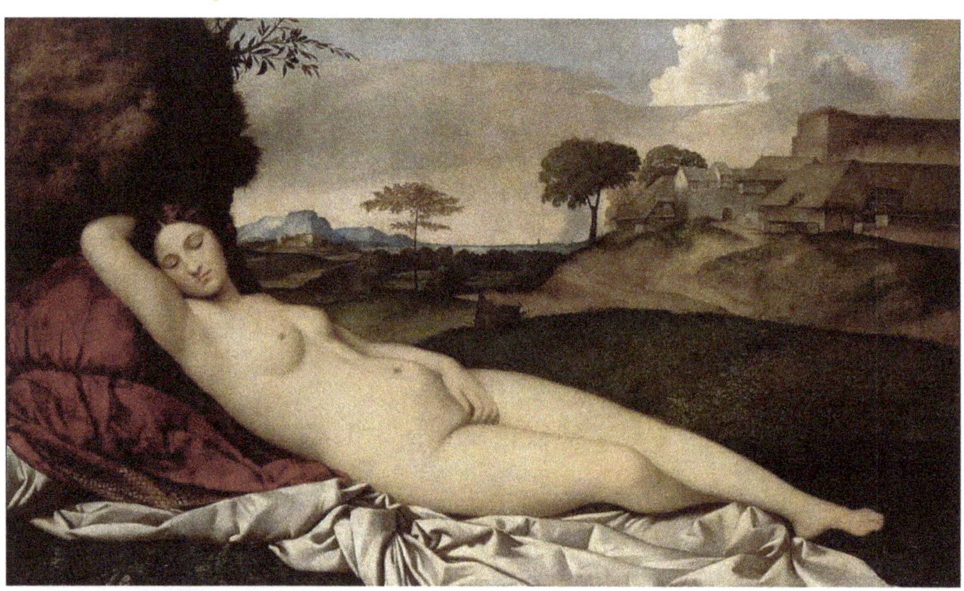

Giorgione, The Sleeping Venus (Venere Dormiente) or Dresden Venus. It was probably completed, after the death of "Big George" at age 33 in 1510, by his contemporary Titian, who lived and painted, including more versions of Venus, to the age of 88. Titian's Venus of Urbino (1534), painted for a cardinal but bought by the Duke of Urbino, is very similar, except for open eyes, lowered right arm, and indoor setting.

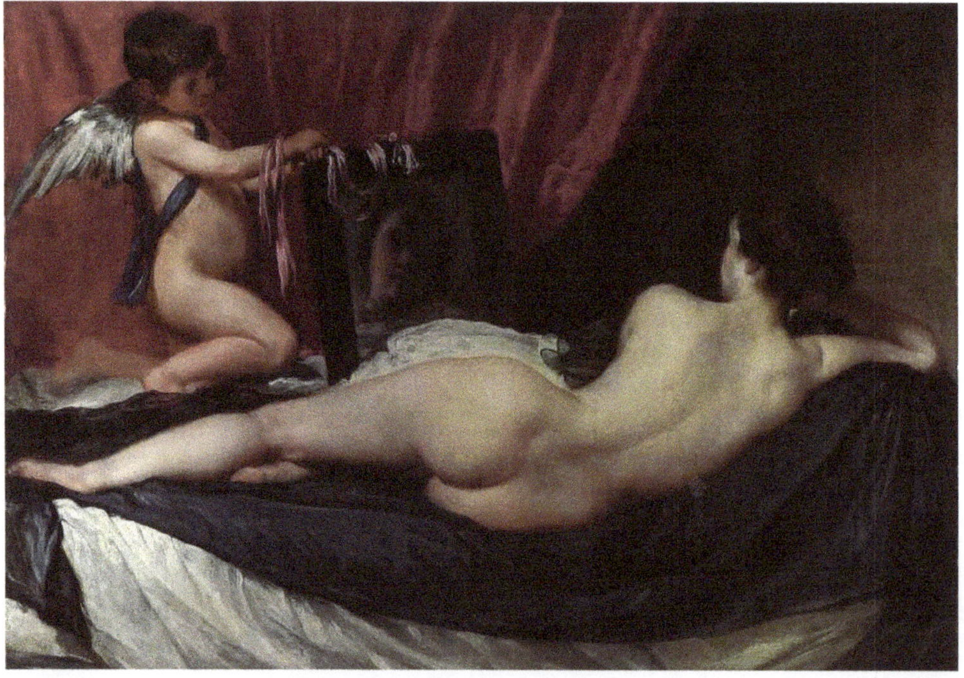

Velázquez, The Rokeby Venus, 1647-1651. It once hung at Rokeby Park in Yorkshire, but has been known by other titles referring to Cupid and the mirror. Velázquez painted no other nudes.

Valentine

Planet Venus has no satellites, but goddess Venus has yet one more. Saint Valentine's Day, which now celebrates love, originally commemorated a martyrdom. There were several martyrs named Valentinus around the third century AD. The usual identification is with Valentine of Rome, of whom legends say that he cured the blindness of the daughter of his jailer, sent her a note from "Your Valentine," and had been imprisoned for administering weddings to Christian soldiers who were forbidden to marry. Chet Raymo rendered the story into a novel, *Valentine* (2005).

Long live whoever loves

This is a graffito found on a wall in Pompeii, the city that was buried by the ash of erupting Vesuvius in 79 AD.

> QUIS[QUIS] AMAT VALEAT, PEREAT QUI NESCIT AMARE.
> BIS TANTO PEREAT QUISQUIS AMARE VETAT.

Valere means "to flourish, be strong." My translation:

> Long live whoever loves! Perish who knows not how!
> Whoever love forbids, twice over perish!

It is scrawled, in handwriting undecipherable for those who know Latin only from our Roman script, as I did till I saw this. Yet it is an elegant couplet in perfect elegiac meter (pause in the middle of the second line). We may speculate whether the fellow who scribbled it had rolled out of a tavern or a brothel, and whether he was quoting a pop ballad or a lost poem of Propertius or Tibullus. I don't think he composed it; he drops a syllable of the first word, and breaks the lines in the wrong places (that, though, was not unusual). Was there a sting of personal feeling behind *vetat*—did he love someone from whom he was vetoed, wrong family, class, age, gender? I described, in a blog post of Valentine's Day 2015, how I puzzled over the scrawl, which I found in a footnote (intended to illustrate that ordinary Romans were literate people) in a small book by Lionel Cassidy, *Libraries in the Ancient World*, sent to me by "Bonnie [who] lies over the ocean."

> . . . Thou hast committed
> Fornication: but that was in another country,
>
> And besides, the wench is dead.
>
> Christopher Marlowe, *The Jew of Malta*, 1589

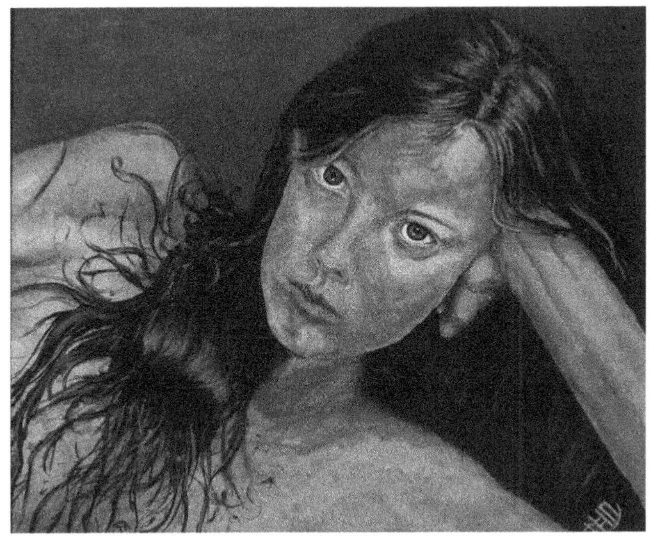

A gift "to Guy & Tilly, my neighbours" (in Lyme Regis) by Tony "Spike" Keegan or Jones; acrylic on cardboard, dated "2007 June 3" (he must have noticed my habit of giving dates in astronomical format). Spike was brought up in a harsh convent school, eked a living with odd jobs, but refused to apply for aid, and died in 2019. He gave us paint-decorated beach pebbles, long poems about his life on the margin, and this expression of the Venus yearning.

Charts

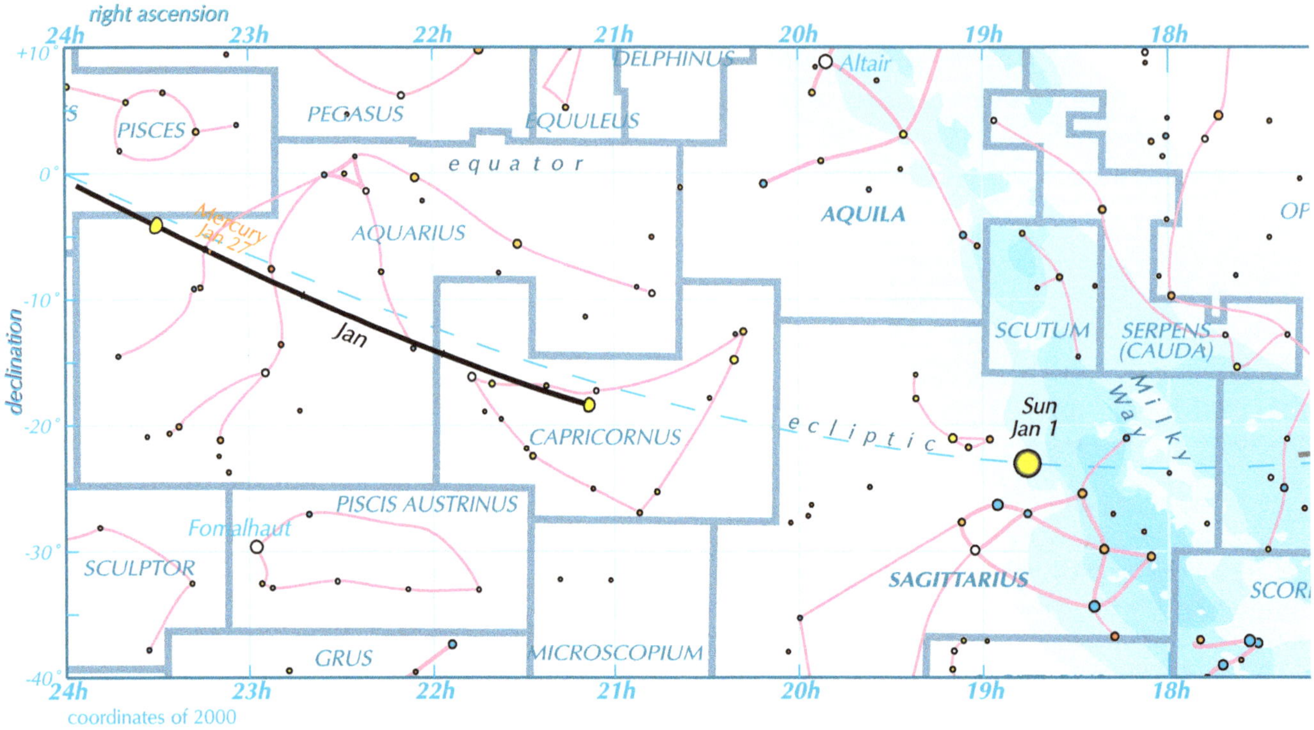

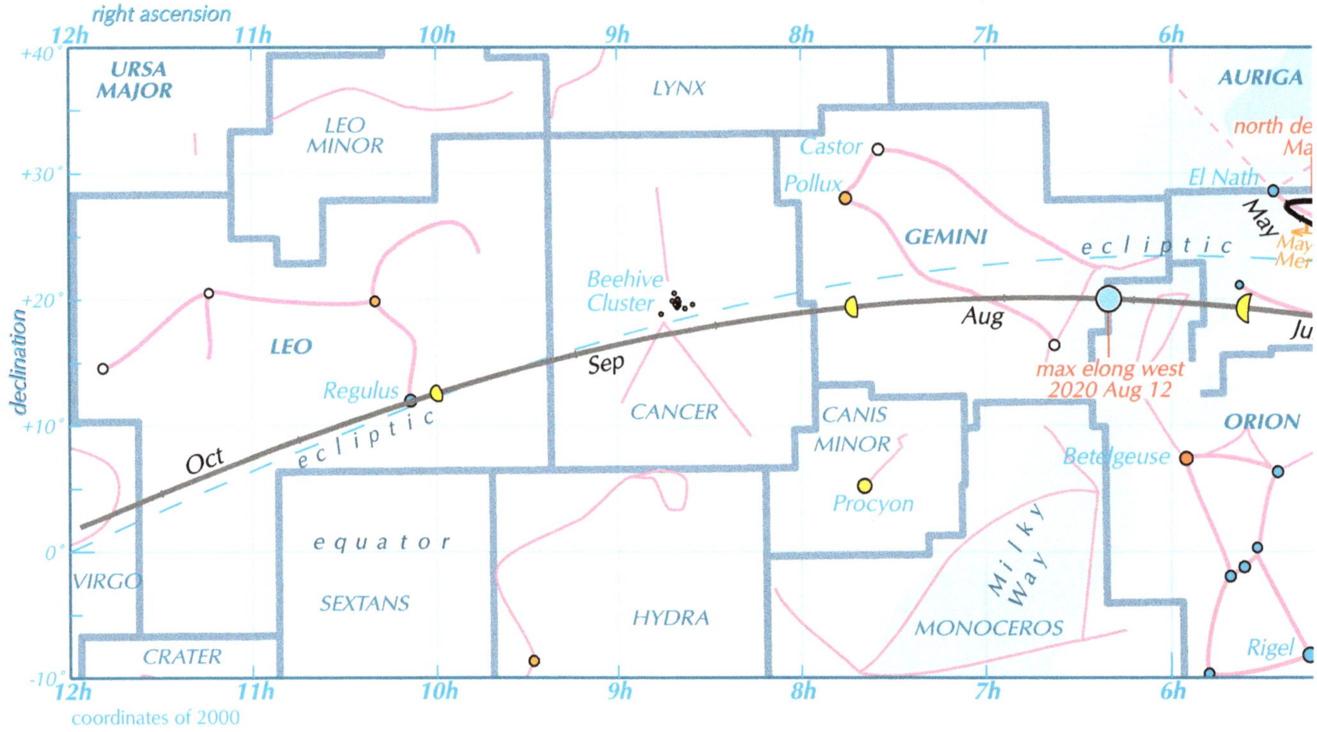

Chart of the path of Venus in eight years (2020 to 2027) and one more (2028) to show how the cycle begins to repeat.

Because the scale is quite large, the charts for a year need two facing pages. The upper chart is for the "left" half of the sky (24h to 12h, or 360° to 180°) and below is the "right" half (12h to 0h, or 180° to 0°). Each has an overlap in the middle.

The charts' framework is equatorial; that is, its dimensions are declination and right ascension. The ecliptic is marked by dashes 2° long. Also shown are the constellation boundaries that were made official in 1930; form-lines suggesting the imaginary shapes of the constellations; and bright stars (colored to represent their spectral classes).

At the 1st of each month, the sunlit part of the planet is drawn, at 300 times its real size. When in the part of its orbit distant from Earth, it is small and almost circular; when near (during the swoop from east to west through inferior conjunction), it is relatively

2020

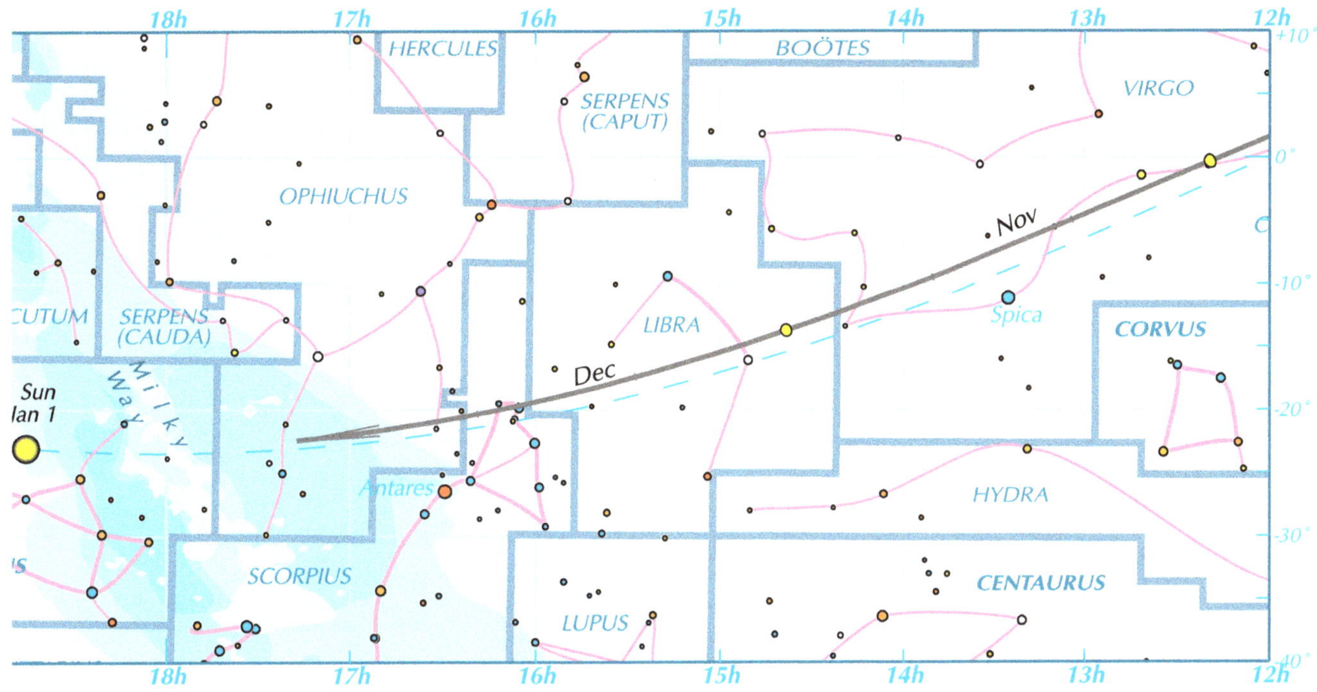

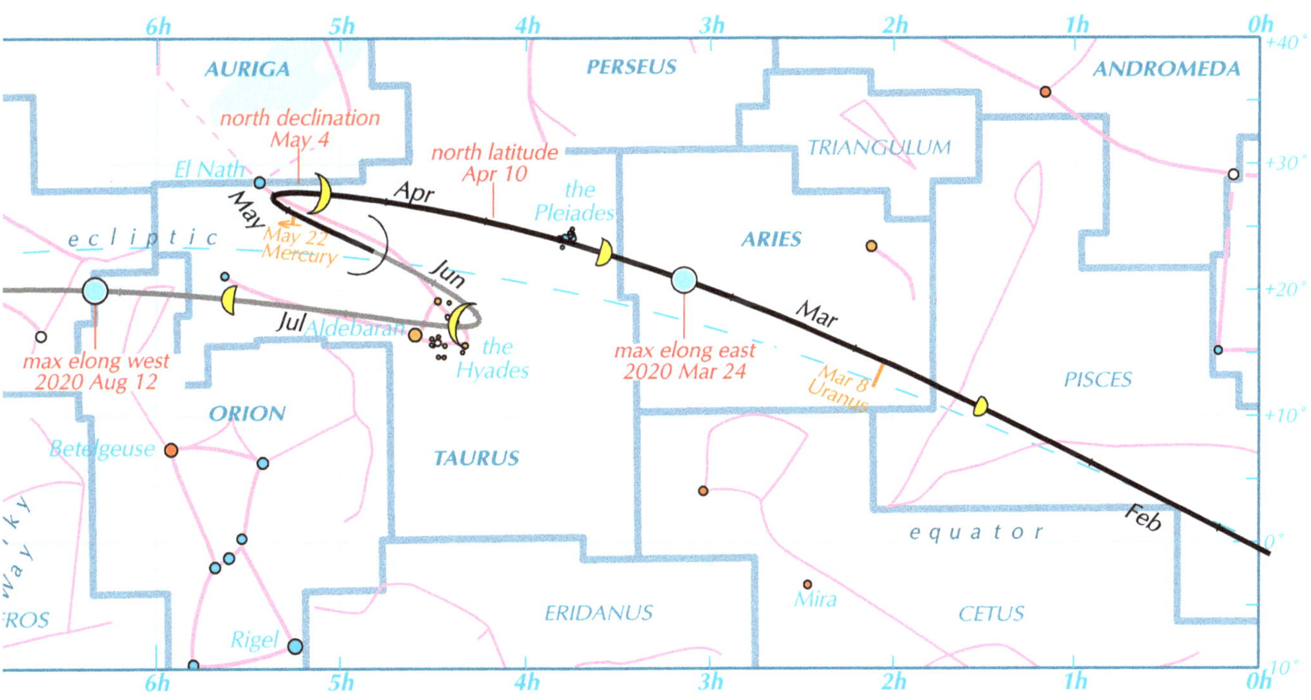

huge though slender, and always facing toward the Sun. Remember that if the planet were drawn at true size it would be a dot at the center of the circle of which the crescent is part.

The path is black when the planet is in the evening sky, gray in the morning sky. So the transition from black to gray indicates the moment of inferior conjunction; from gray to black, superior conjunction.

Ticks on the path mark the position at the 11th and 21st of the month.

Venus is always relatively near to the Sun, so its path for the year has to start and end near the Sun's January 1 position. But in some years, such as 2020, it starts on the Sun's east and draws away into the evening sky; other Januaries, such as 2021, find it on the Sun's west, so that it begins the year by falling toward the sunrise and inferior conjunction.

At instants when Venus is closest to another planet (appulses), a line connects the two, and a short part of the other planet's path is drawn.

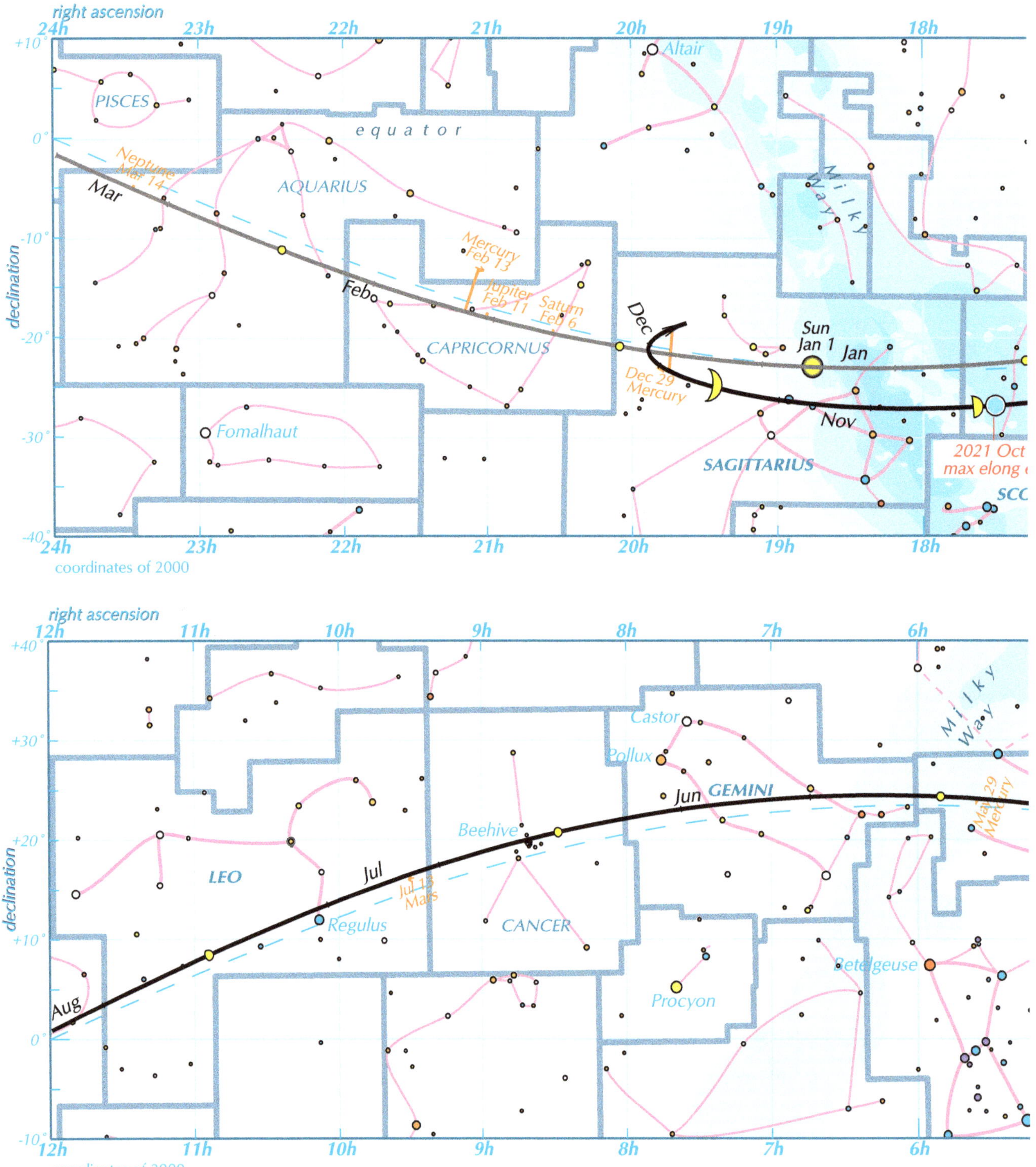
right ascension
24h
23h
22h
21h
20h
19h
18h
+10°
0°
-10°
-20°
-30°
-40°
declination
Altair
PISCES
equator
Milky Way
Neptune Mar 14
Mar
AQUARIUS
Mercury Feb 13
Feb
Jupiter Feb 11
Saturn Feb 6
Dec
Sun Jan 1
Jan
CAPRICORNUS
Dec 29 Mercury
Nov
Fomalhaut
SAGITTARIUS
2021 Oct
max elong
coordinates of 2000
right ascension
12h
11h
10h
9h
8h
7h
6h
+40°
+30°
+20°
+10°
0°
-10°
declination
Milky Way
Castor
Pollux
Jun
GEMINI
May 29 Mercury
Beehive
LEO
Jul
Jul 13 Mars
CANCER
Regulus
Betelgeuse
Aug
Procyon
coordinates of 2000

2021

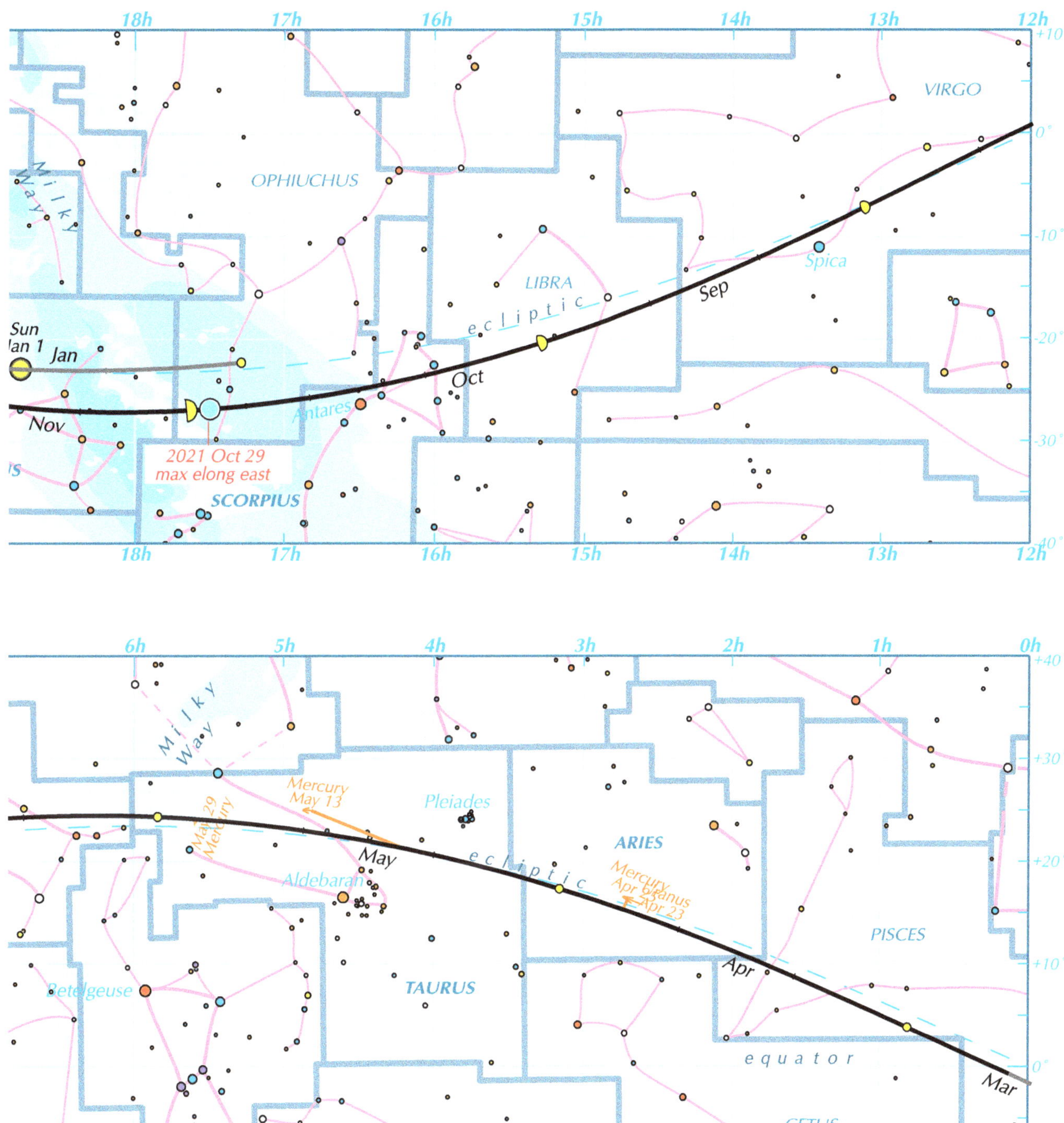

18h
17h
16h
15h
14h
13h
12h
+10°
0°
-10°
-20°
-30°
-40°
VIRGO
OPHIUCHUS
Milky Way
LIBRA
Spica
ecliptic
Sep
Sun
Jan 1
Jan
Oct
Nov
Antares
2021 Oct 29
max elong east
SCORPIUS
6h
5h
4h
3h
2h
1h
0h
+40°
+30°
+20°
+10°
-10°
Milky Way
Mercury
May 13
May 29
Mercury
Pleiades
May
ARIES
Aldebaran
Mercury
Apr 6
Uranus
Apr 23
PISCES
TAURUS
Apr
Betelgeuse
equator
Mar
CETUS
Rigel

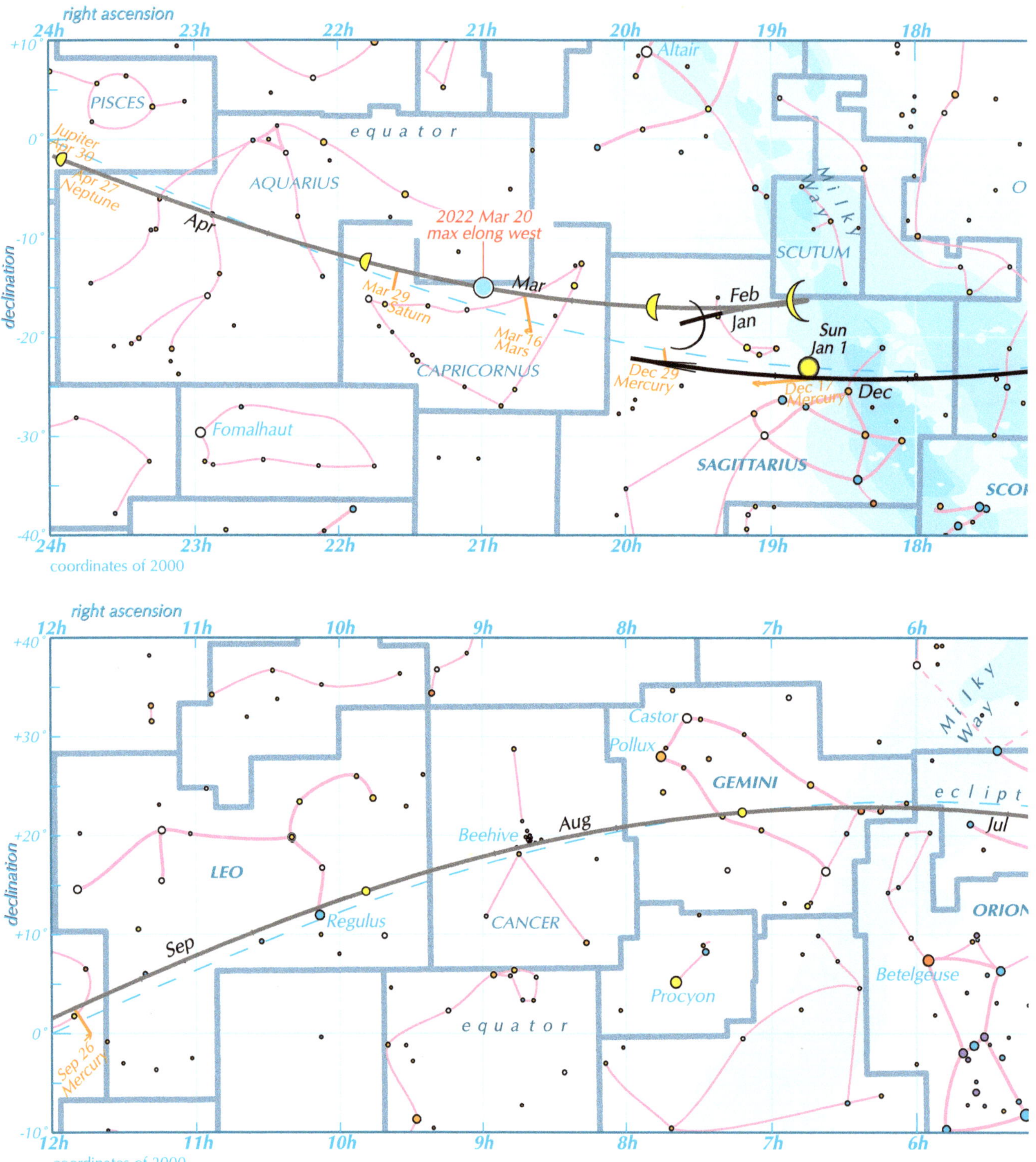
right ascension
24h
23h
22h
21h
20h
19h
18h
+10°
0°
-10°
-20°
-30°
-40°
declination
Altair
PISCES
equator
AQUARIUS
Jupiter Apr 30
Apr 27 Neptune
Apr
2022 Mar 20
max elong west
Mar
Feb
Jan
Mar 29 Saturn
Mar 16 Mars
SCUTUM
Milky Way
Sun Jan 1
Dec
Dec 29 Mercury
Dec 17 Mercury
CAPRICORNUS
Fomalhaut
SAGITTARIUS
SCO
coordinates of 2000
right ascension
12h
11h
10h
9h
8h
7h
6h
+40°
+30°
+20°
+10°
0°
-10°
Castor
Pollux
GEMINI
ecliptic
Milky Way
Jul
Aug
Beehive
LEO
CANCER
Regulus
ORION
Sep
Betelgeuse
Procyon
equator
Sep 26 Mercury
coordinates of 2000

2022

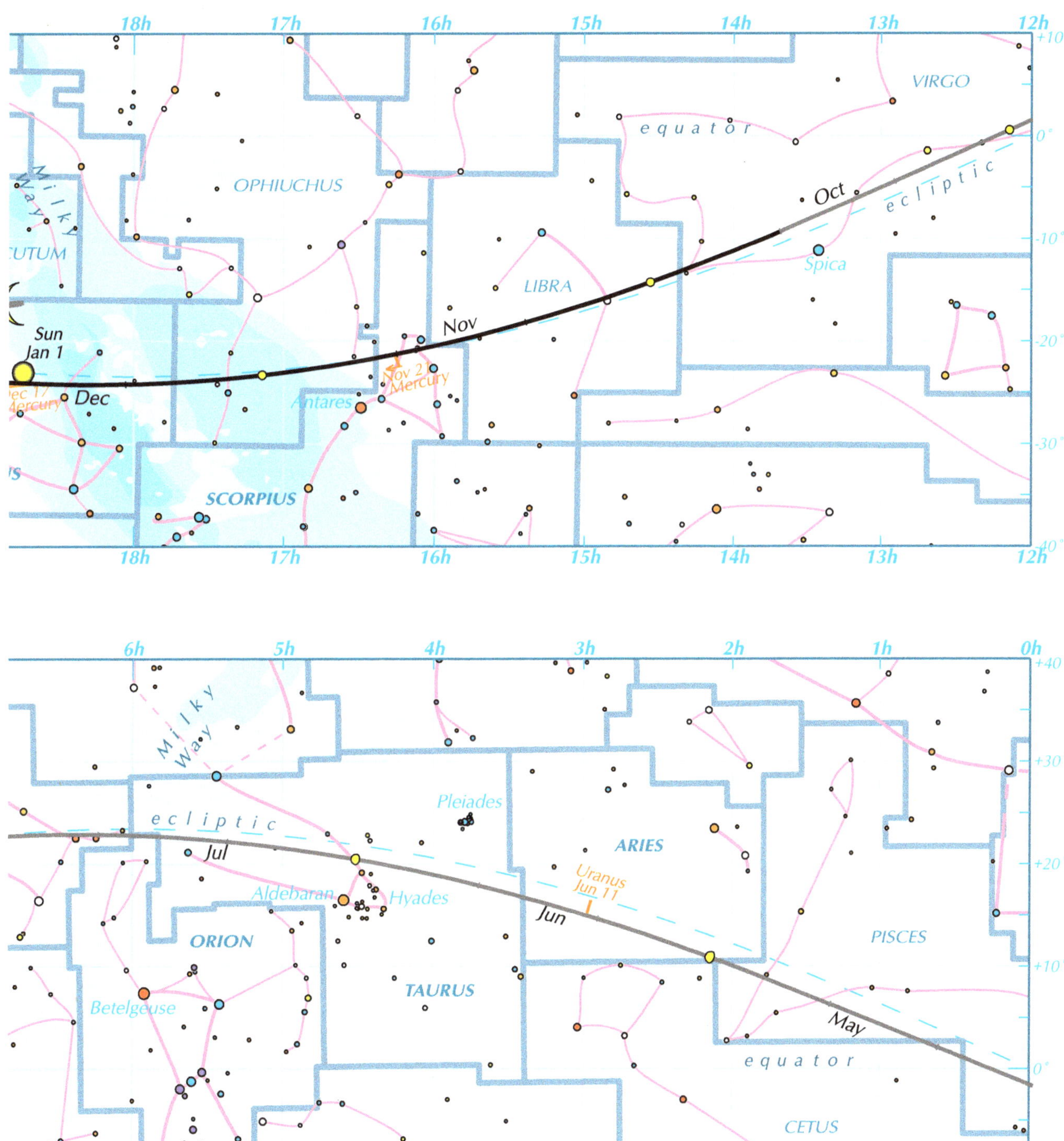
18h
17h
16h
15h
14h
13h
12h
+10°
0°
-10°
-20°
-30°
VIRGO
equator
OPHIUCHUS
Milky Way
Oct
ecliptic
LIBRA
Spica
Nov
Sun
Jan 1
Nov 21
Mercury
Dec
Antares
SCORPIUS
6h
5h
4h
3h
2h
1h
0h
+40°
+30°
+20°
+10°
0°
Milky Way
ecliptic
Pleiades
ARIES
Jul
Uranus
Jun 11
Aldebaran
Hyades
Jun
ORION
PISCES
TAURUS
Betelgeuse
May
equator
CETUS
Rigel

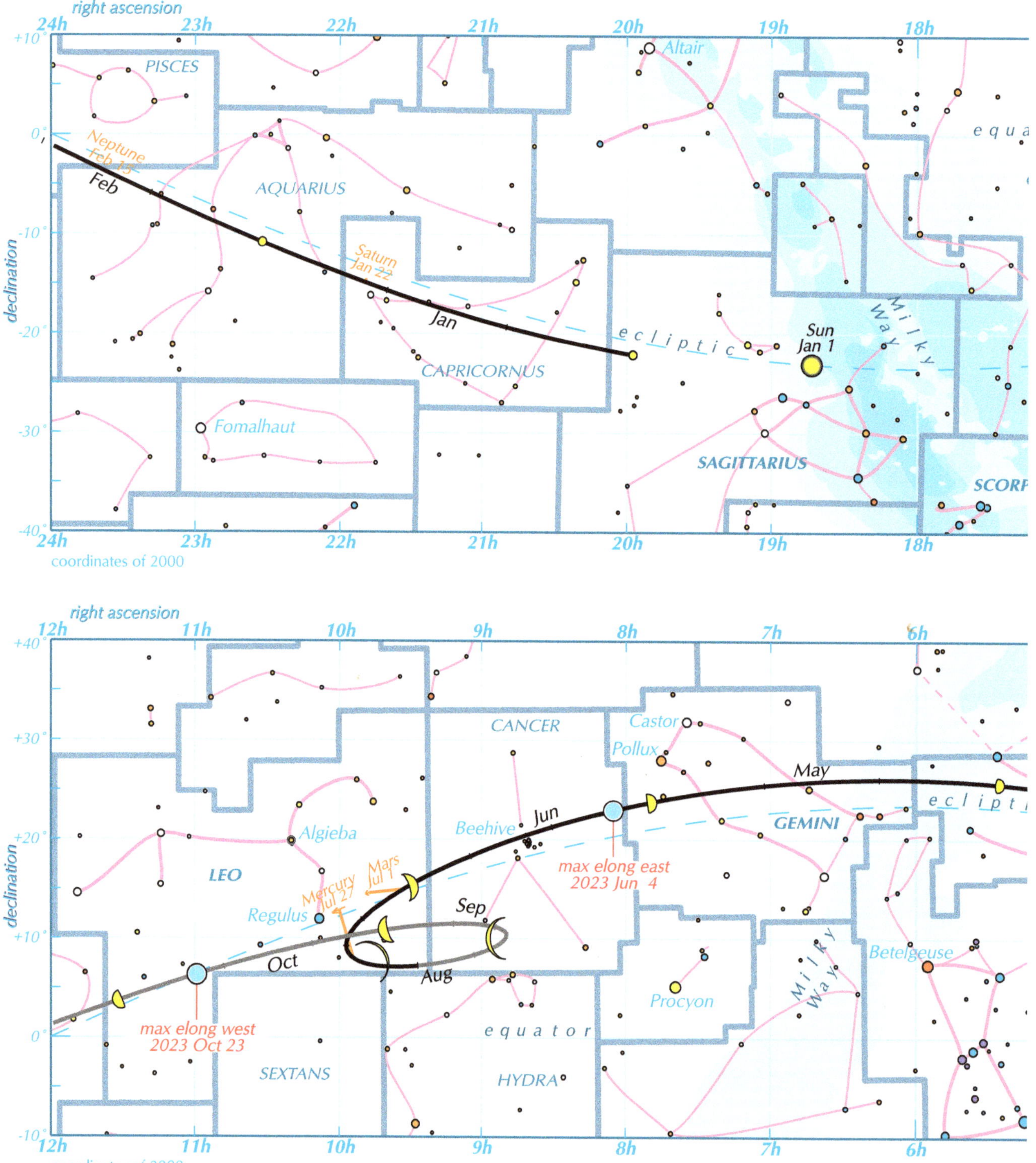

right ascension
24h
23h
22h
21h
20h
19h
18h
PISCES
Altair
AQUARIUS
Neptune Feb 15
Feb
Saturn Jan 22
Jan
ecliptic
Sun Jan 1
Milky Way
CAPRICORNUS
Fomalhaut
SAGITTARIUS
SCORP
equa
declination
coordinates of 2000
right ascension
12h
11h
10h
9h
8h
7h
6h
CANCER
Castor
Pollux
May
Jun
Beehive
GEMINI
ecliptic
max elong east 2023 Jun 4
Algieba
LEO
Mars Jul 1
Mercury Jul 27
Regulus
Sep
Oct
Aug
Betelgeuse
Procyon
Milky Way
max elong west 2023 Oct 23
equator
SEXTANS
HYDRA
declination
coordinates of 2000

2023

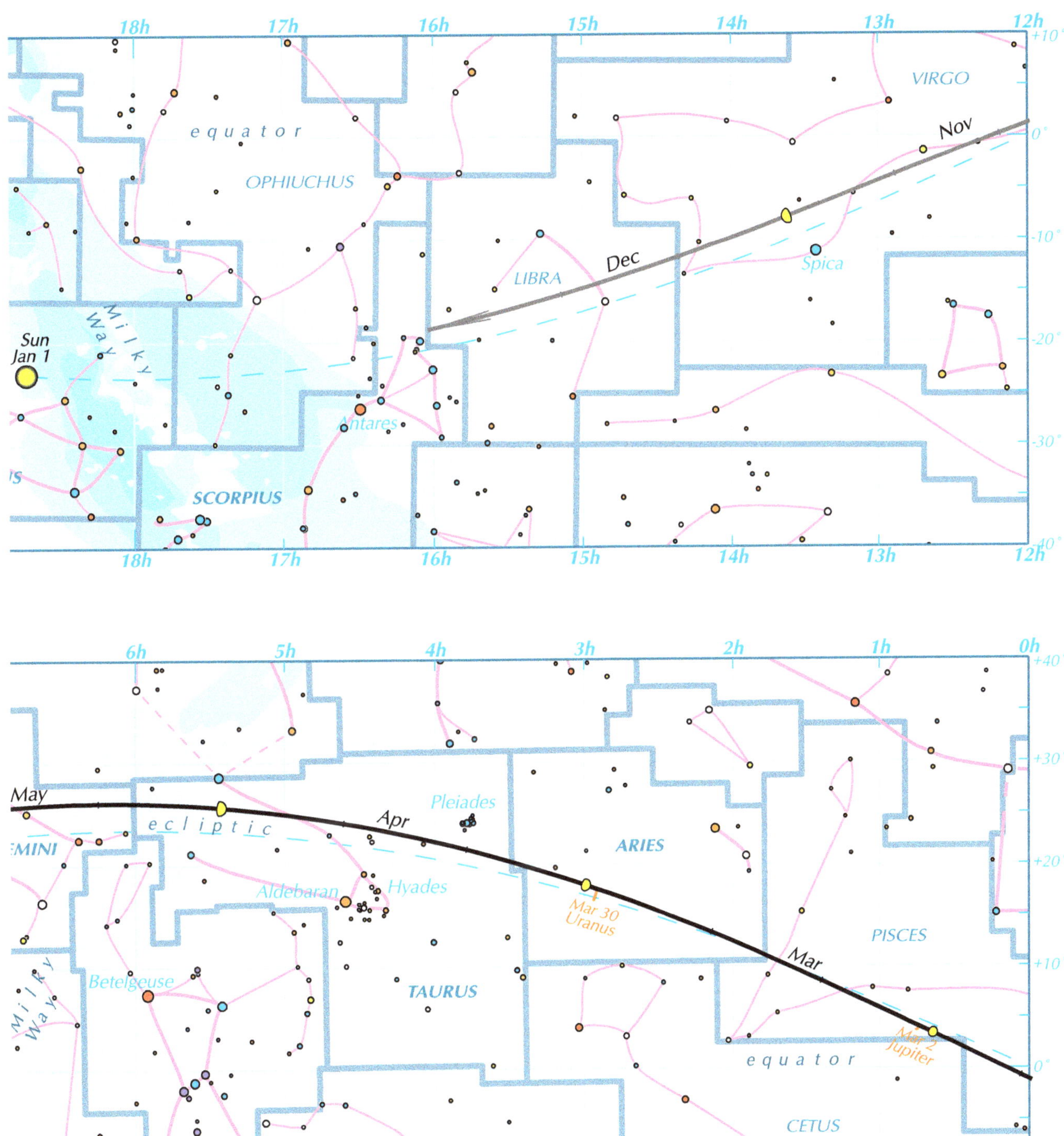
18h
17h
16h
15h
14h
13h
12h
+10°
0°
-10°
-20°
-30°
-40°
VIRGO
equator
OPHIUCHUS
Nov
Dec
LIBRA
Spica
Milky Way
Sun
Jan 1
Antares
SCORPIUS
6h
5h
4h
3h
2h
1h
0h
+40°
+30°
+20°
+10°
0°
-10°
May
ecliptic
Apr
Pleiades
ARIES
EMINI
Aldebaran
Hyades
Mar 30
Uranus
PISCES
Mar
Betelgeuse
TAURUS
Milky Way
Mar 2
Jupiter
equator
CETUS
Rigel

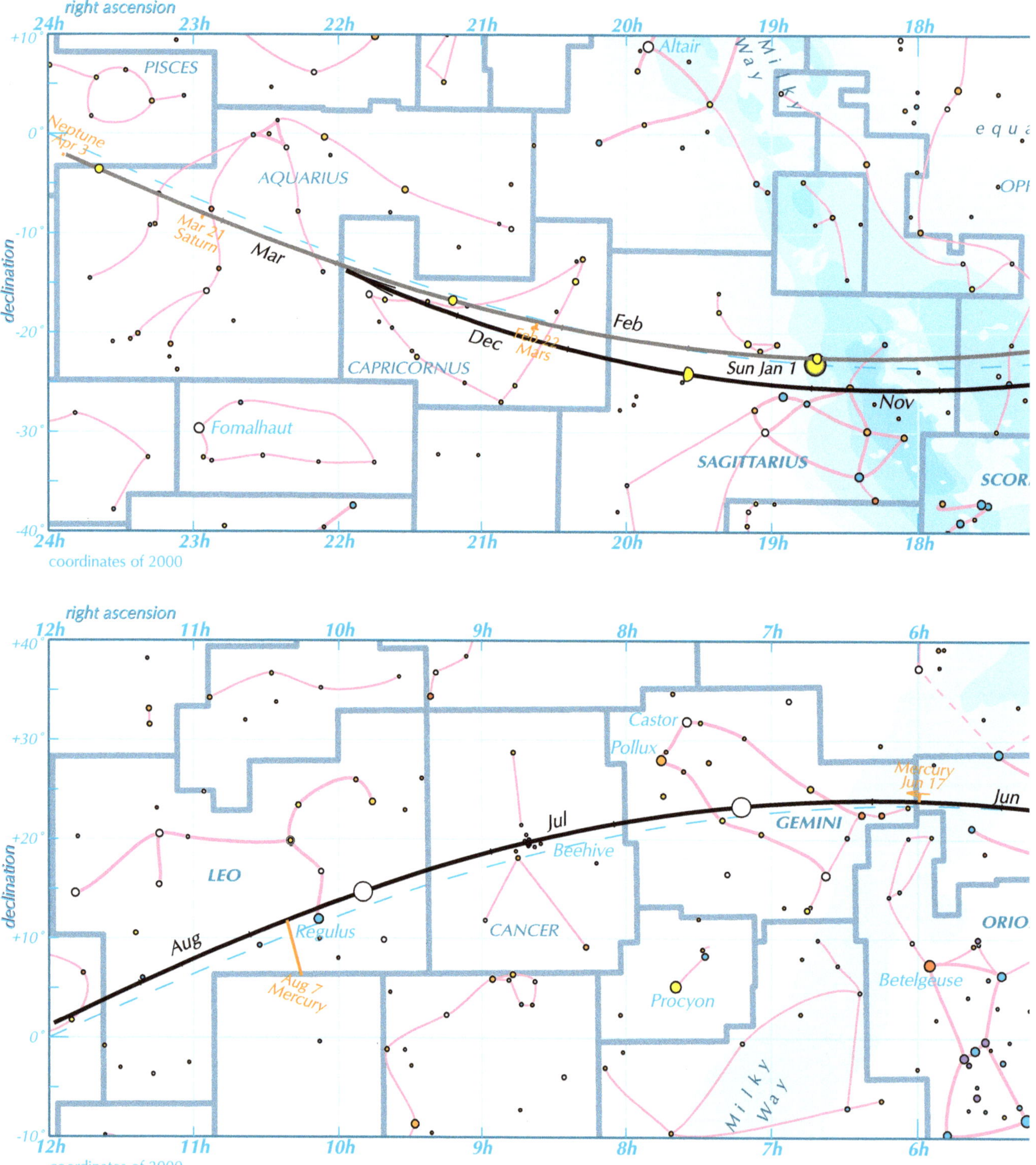

right ascension
24h
23h
22h
21h
20h
19h
18h
+10°
0°
-10°
-20°
-30°
-40°
declination
PISCES
Altair
Milky Way
Neptune Apr 3
AQUARIUS
Mar 21 Saturn
Mar
Feb
Dec
Feb 22 Mars
CAPRICORNUS
Sun Jan 1
Nov
Fomalhaut
SAGITTARIUS
SCOR
OPH
equa
coordinates of 2000
right ascension
12h
11h
10h
9h
8h
7h
6h
+40°
+30°
+20°
+10°
0°
-10°
declination
Castor
Pollux
Mercury Jun 17
Jun
Jul
GEMINI
Beehive
LEO
CANCER
Aug
Regulus
Aug 7 Mercury
ORIO
Betelgeuse
Procyon
Milky Way
coordinates of 2000

2024

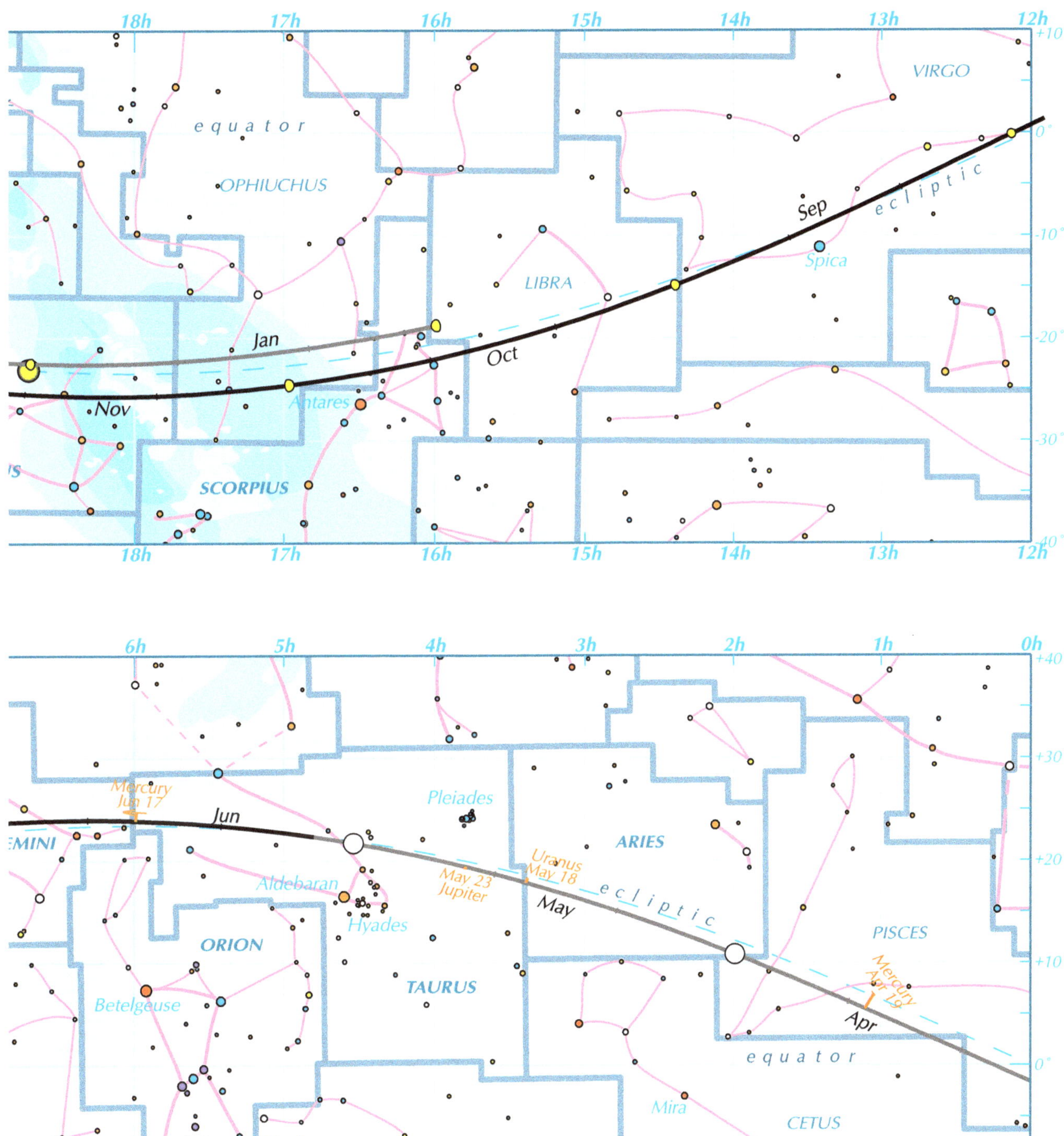
18h
17h
16h
15h
14h
13h
12h
+10°
0°
-10°
-20°
-30°
-40°
VIRGO
equator
OPHIUCHUS
ecliptic
Sep
Spica
LIBRA
Jan
Oct
Nov
Antares
SCORPIUS
6h
5h
4h
3h
2h
1h
0h
+40°
+30°
+20°
+10°
-10°
Mercury
Jun 17
Jun
EMINI
Pleiades
ARIES
Uranus
May 18
May 23
Jupiter
Aldebaran
May
Hyades
ORION
PISCES
TAURUS
Mercury
Apr 19
Betelgeuse
Apr
equator
Mira
CETUS
Rigel

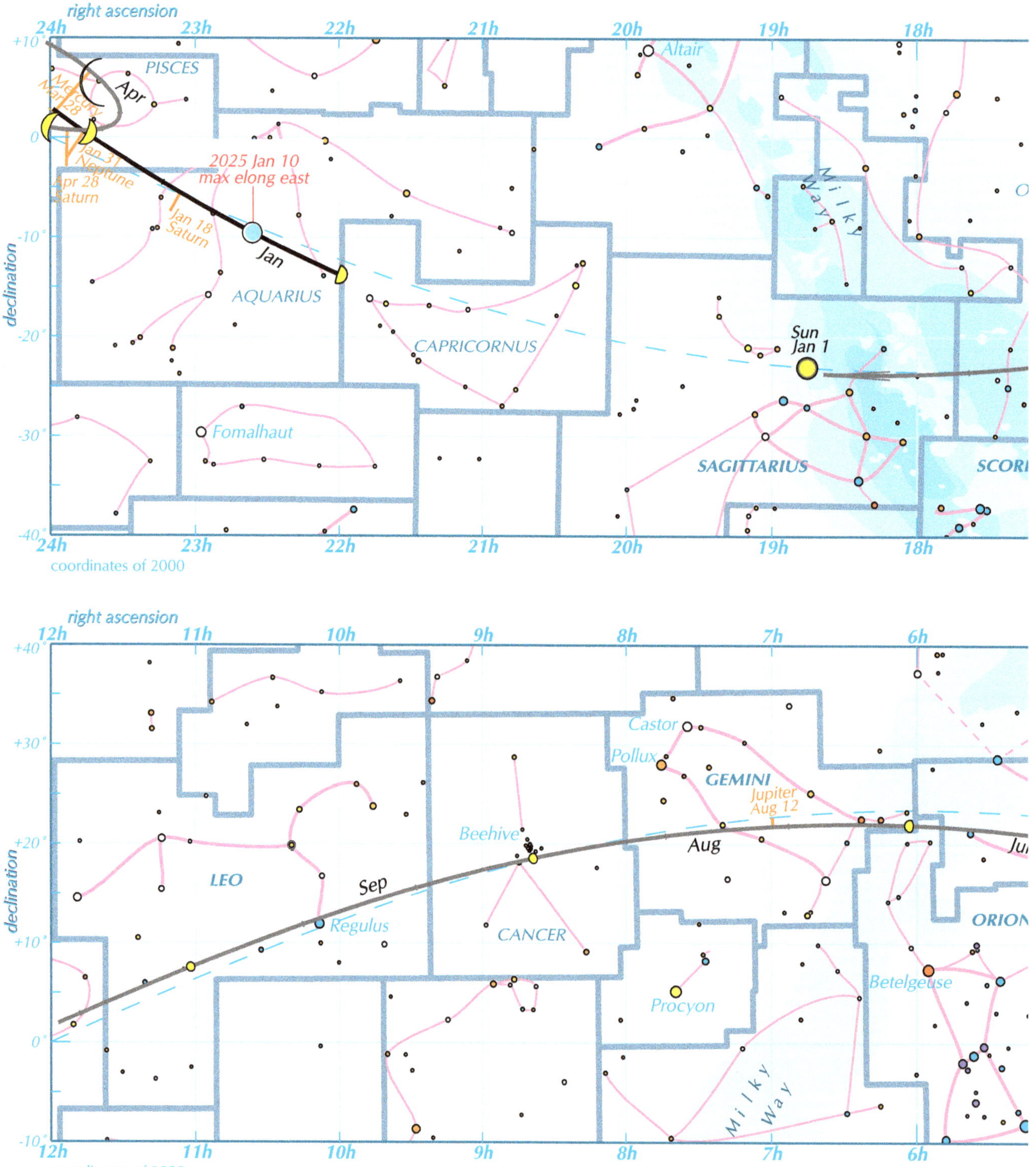
right ascension
PISCES
Apr
Mercury Mar 28
Jan 31 Neptune
Apr 28 Saturn
2025 Jan 10
max elong east
Jan 18 Saturn
Jan
AQUARIUS
CAPRICORNUS
Altair
Milky Way
Sun
Jan 1
SAGITTARIUS
Fomalhaut
declination
coordinates of 2000
right ascension
Castor
Pollux
GEMINI
Jupiter
Aug 12
Beehive
Aug
Sep
LEO
Regulus
CANCER
ORION
Betelgeuse
Procyon
Milky Way
declination
coordinates of 2000

2025

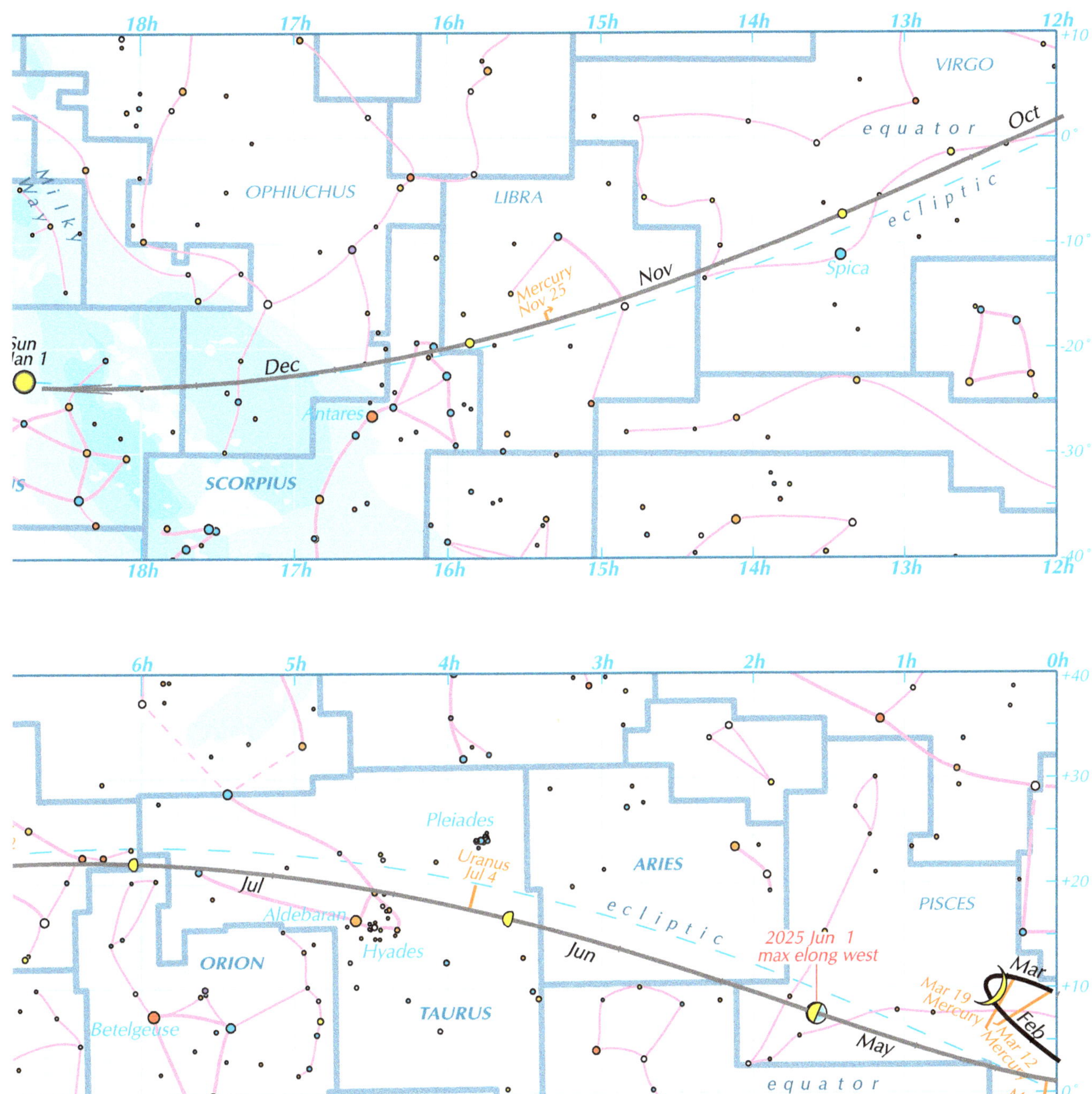

18h
17h
16h
15h
14h
13h
12h
+10°
0°
-10°
-20°
-30°
-40°
VIRGO
equator
Oct
ecliptic
OPHIUCHUS
LIBRA
Milky Way
Spica
Mercury Nov 25
Nov
Sun Jan 1
Dec
Antares
SCORPIUS
6h
5h
4h
3h
2h
1h
0h
+40°
+30°
+20°
+10°
Pleiades
ARIES
Uranus Jul 4
Jul
Aldebaran
ecliptic
PISCES
Hyades
Jun
2025 Jun 1 max elong west
ORION
Mar
Mar 19 Mercury
TAURUS
Feb
Betelgeuse
Mar 12 Mercury
May
equator
May 3 Neptune
CETUS
Rigel

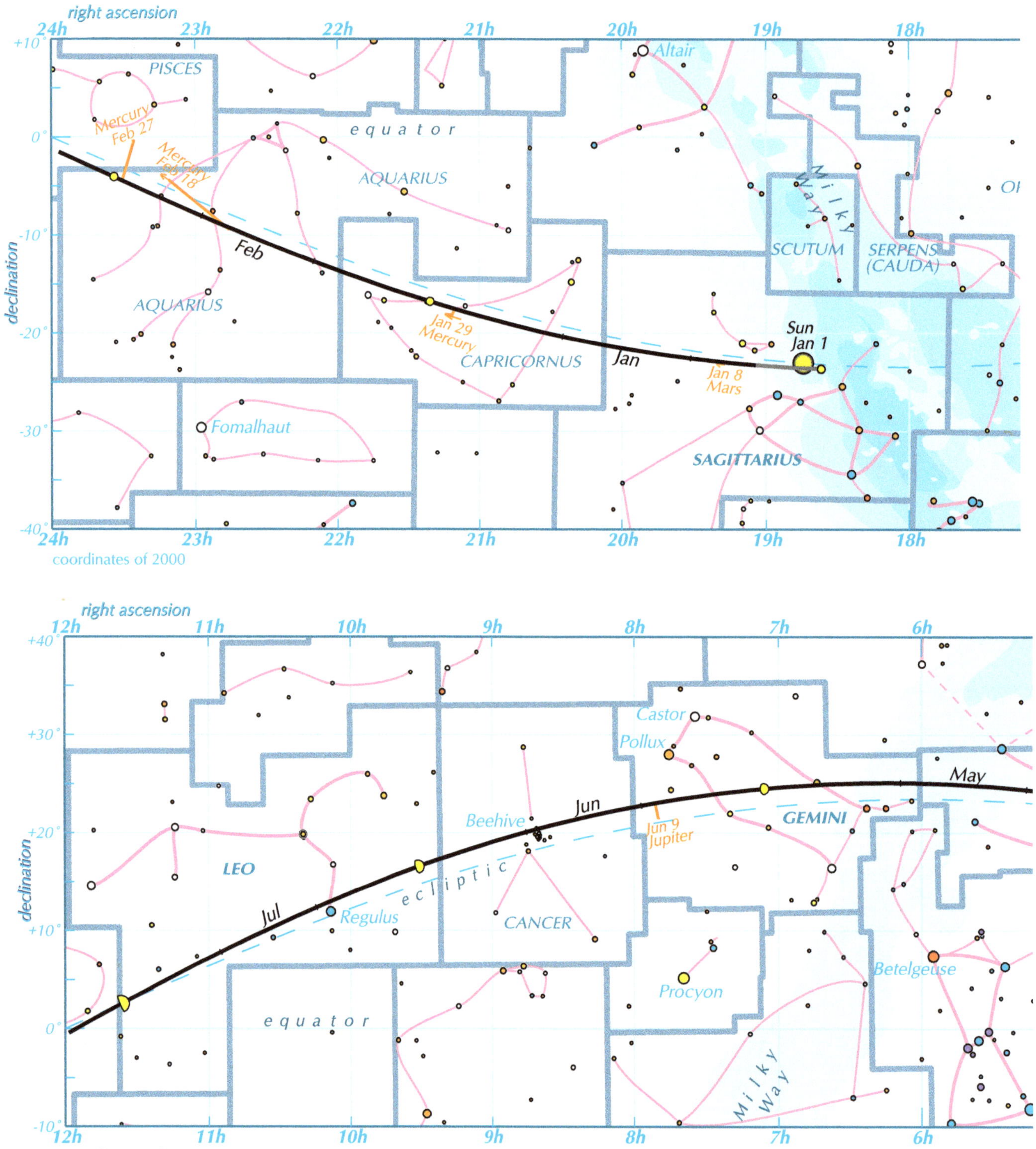
right ascension
24h
23h
22h
21h
20h
19h
18h
+10°
0°
-10°
-20°
-30°
-40°
declination
PISCES
Altair
Mercury Feb 27
Mercury Feb 18
equator
AQUARIUS
Milky Way
SCUTUM
SERPENS (CAUDA)
Feb
AQUARIUS
Jan 29 Mercury
Sun Jan 1
CAPRICORNUS
Jan
Jan 8 Mars
Fomalhaut
SAGITTARIUS
coordinates of 2000
right ascension
12h
11h
10h
9h
8h
7h
6h
+40°
+30°
+20°
+10°
0°
-10°
declination
Castor
Pollux
May
Beehive
Jun
Jun 9 Jupiter
GEMINI
LEO
ecliptic
Jul
Regulus
CANCER
Betelgeuse
Procyon
equator
Milky Way
coordinates of 2000

2026

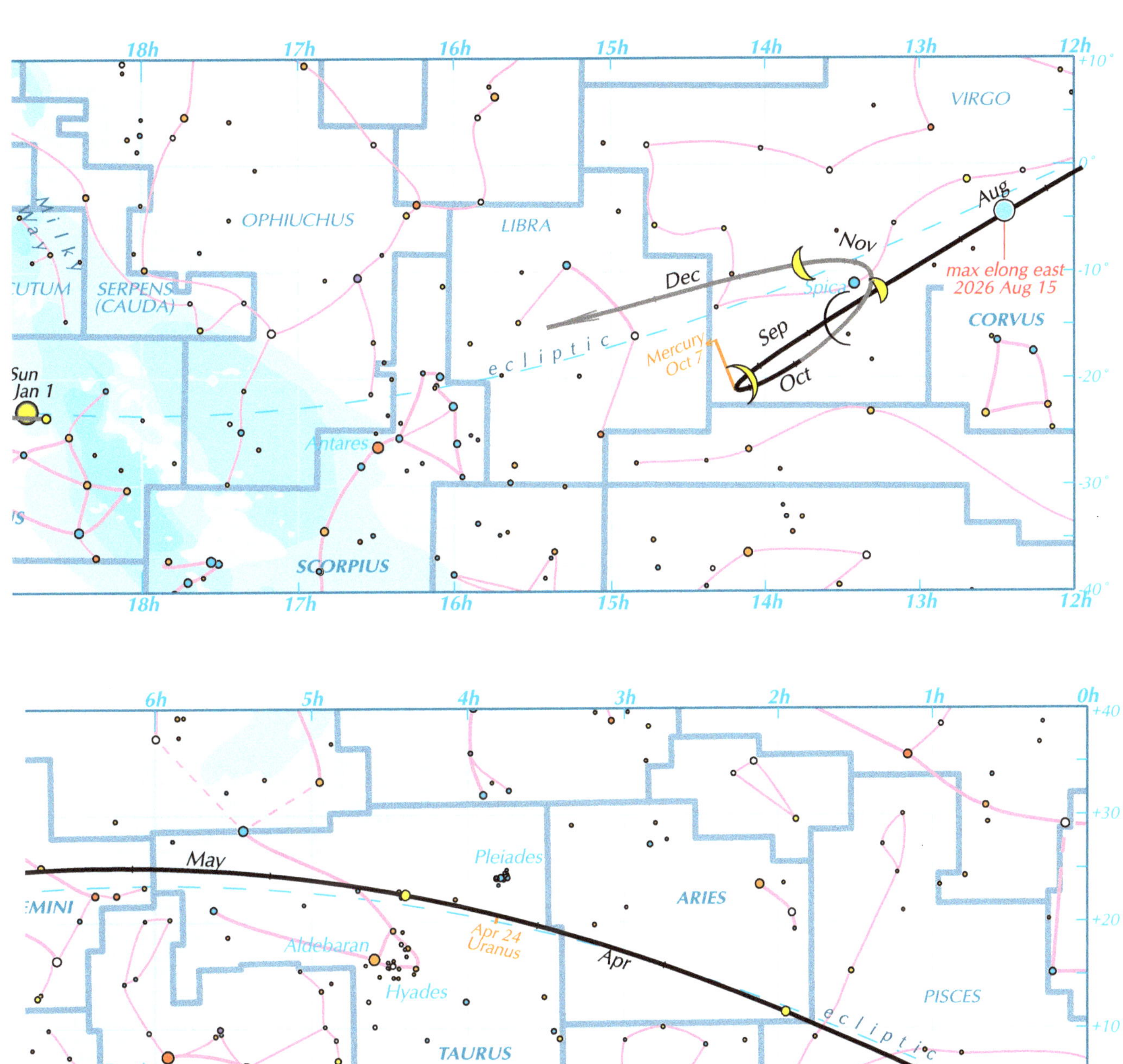

VIRGO
OPHIUCHUS
LIBRA
Milky Way
SCUTUM
SERPENS (CAUDA)
Sun Jan 1
Antares
SCORPIUS
CORVUS
Spica
Aug
Nov
Dec
Sep
Oct
ecliptic
Mercury Oct 7
max elong east 2026 Aug 15
May
Apr
Mar
GEMINI
Pleiades
Aldebaran
Hyades
Apr 24 Uranus
ARIES
PISCES
TAURUS
Betelgeuse
Rigel
equator
CETUS
Mar 8 Saturn
Mar 7 Neptune

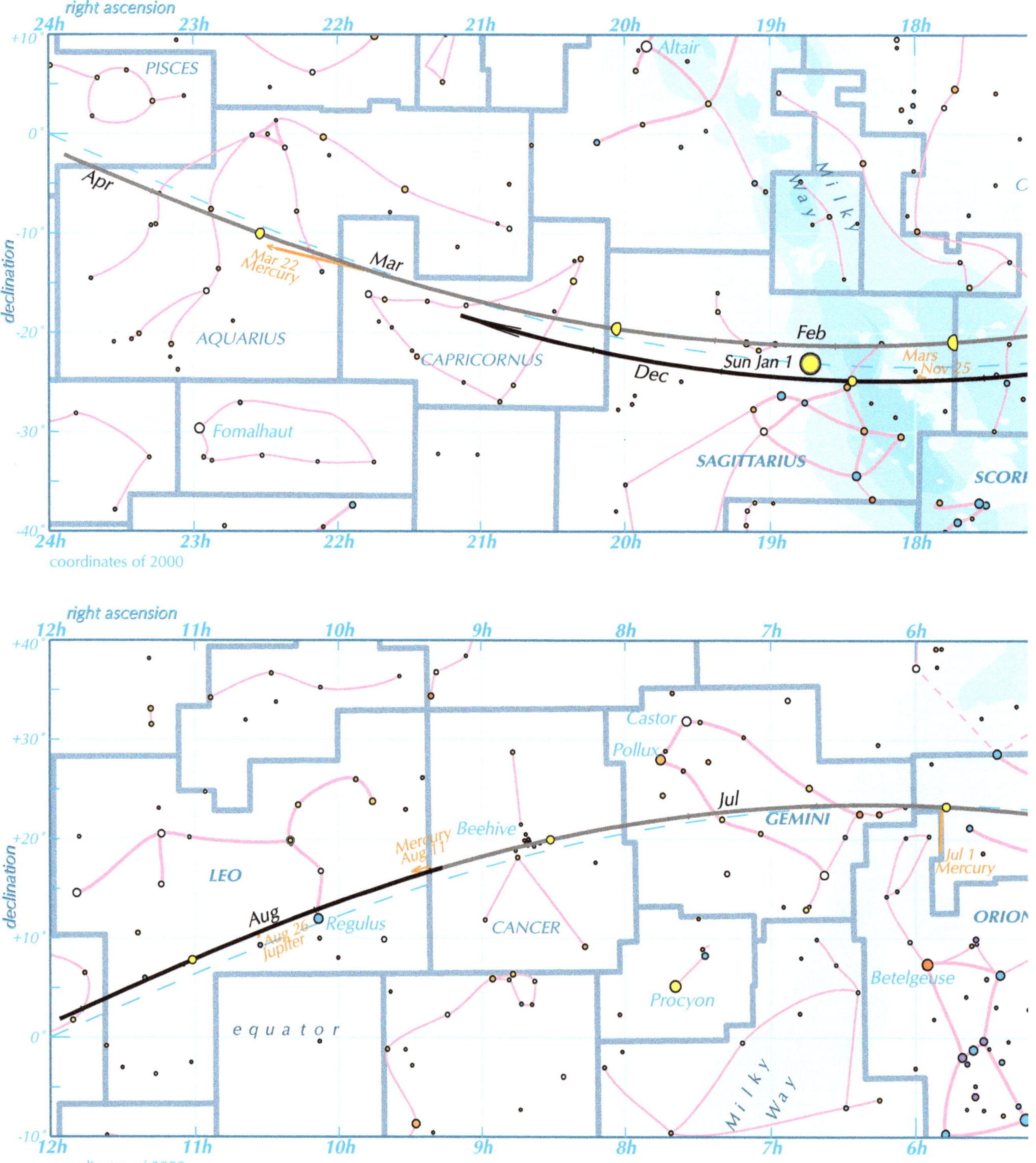
right ascension
24h
23h
22h
21h
20h
19h
18h
+10°
0°
-10°
-20°
-30°
-40°
declination
PISCES
Altair
Apr
Mar
Mar 22
Mercury
Milky Way
AQUARIUS
CAPRICORNUS
Feb
Sun Jan 1
Dec
Mars
Nov 25
Fomalhaut
SAGITTARIUS
SCORI
coordinates of 2000
right ascension
12h
11h
10h
9h
8h
7h
6h
+40°
+30°
+20°
+10°
0°
-10°
declination
Castor
Pollux
Jul
GEMINI
Beehive
Mercury
Aug 11
LEO
Jul 1
Mercury
Aug
Aug 26
Jupiter
Regulus
CANCER
ORION
Betelgeuse
Procyon
equator
Milky Way
coordinates of 2000

2027

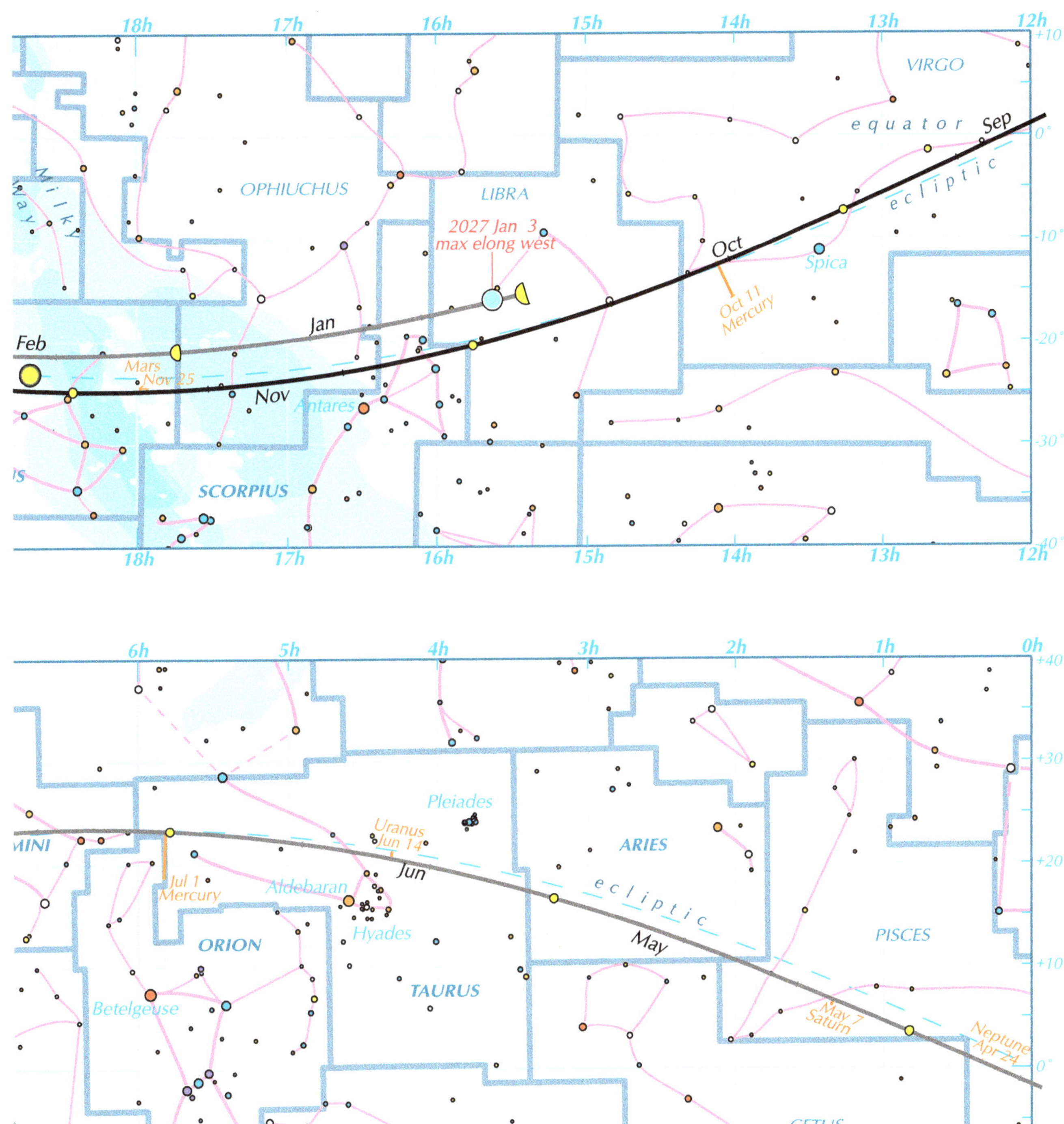
18h
17h
16h
15h
14h
13h
12h
+10°
0°
-10°
-20°
-30°
-40°
VIRGO
equator
Sep
ecliptic
OPHIUCHUS
LIBRA
Milky Way
2027 Jan 3
max elong west
Oct
Spica
Oct 11
Mercury
Jan
Feb
Mars
Nov 25
Nov
Antares
SCORPIUS
6h
5h
4h
3h
2h
1h
0h
+40°
+30°
+20°
+10°
0°
-10°
Pleiades
Uranus
Jun 14
Jun
ARIES
MINI
Jul 1
Mercury
Aldebaran
ecliptic
Hyades
ORION
May
PISCES
TAURUS
Betelgeuse
May 7
Saturn
Neptune
Apr 24
CETUS
Rigel

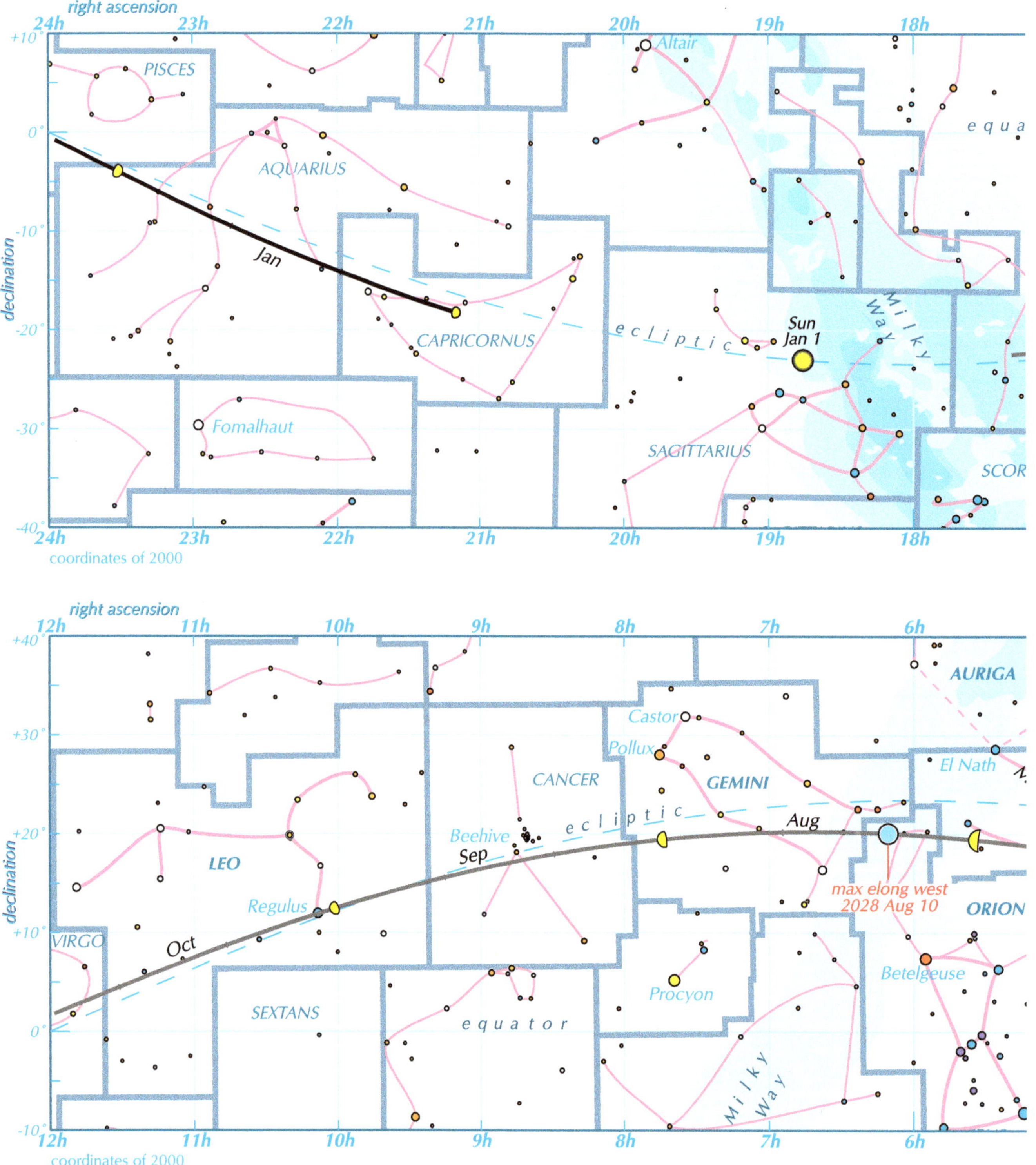
right ascension
24h
23h
22h
21h
20h
19h
18h
+10°
0°
-10°
-20°
-30°
-40°
declination
Altair
PISCES
AQUARIUS
Jan
CAPRICORNUS
ecliptic
Sun
Jan 1
Milky Way
equa
Fomalhaut
SAGITTARIUS
SCOR
coordinates of 2000
right ascension
12h
11h
10h
9h
8h
7h
6h
+40°
+30°
+20°
+10°
0°
-10°
declination
AURIGA
Castor
Pollux
El Nath
CANCER
GEMINI
Beehive
ecliptic
Aug
Sep
LEO
max elong west
2028 Aug 10
Regulus
ORION
VIRGO
Oct
Betelgeuse
Procyon
SEXTANS
equator
Milky Way
coordinates of 2000

2028

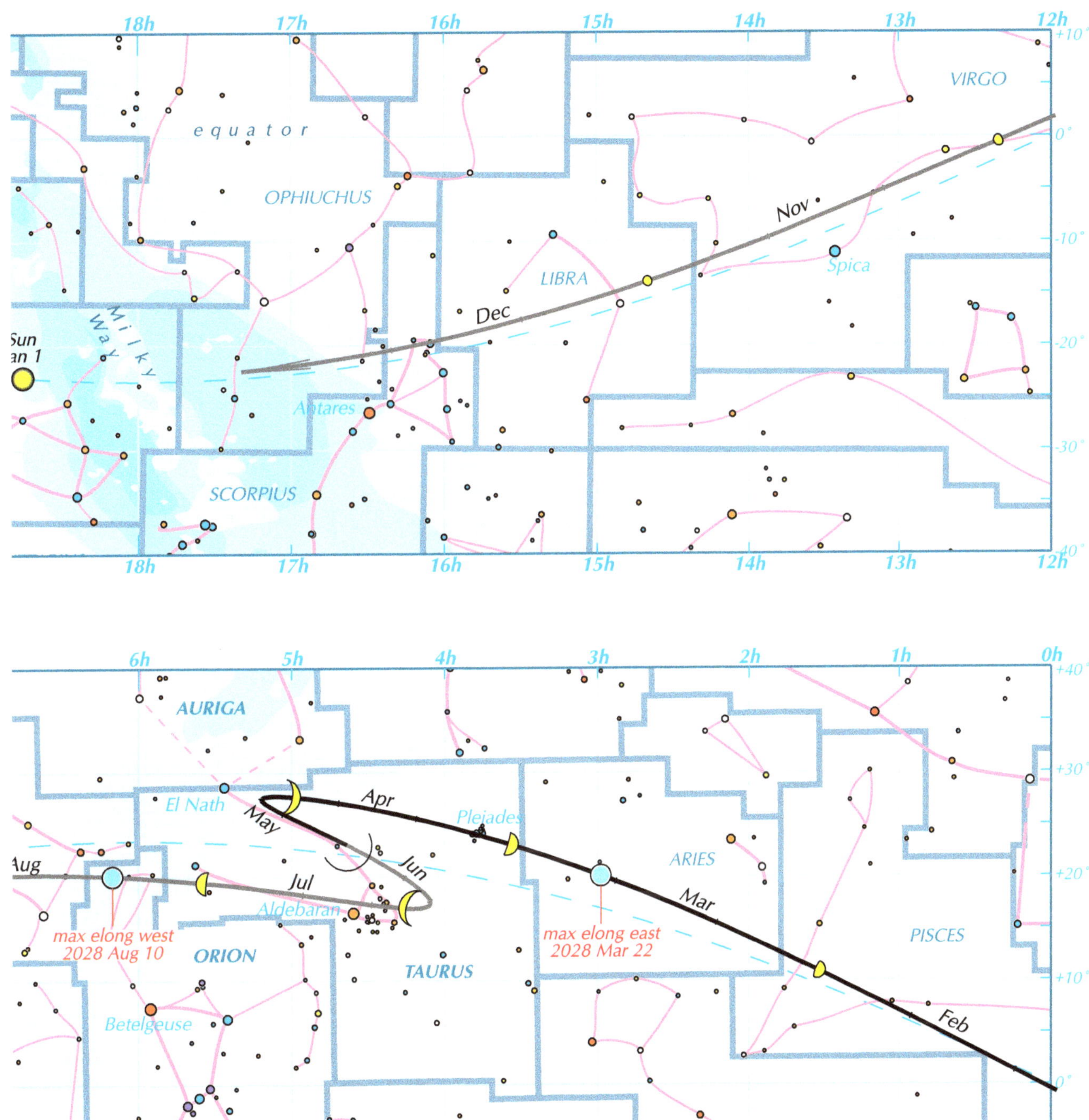

18h
17h
16h
15h
14h
13h
12h
+10°
VIRGO
equator
0°
OPHIUCHUS
Nov
-10°
Spica
LIBRA
Dec
Milky Way
Sun
Jan 1
-20°
Antares
-30°
SCORPIUS
-40°
6h
5h
4h
3h
2h
1h
0h
+40°
AURIGA
+30°
El Nath
Apr
May
Pleiades
ARIES
Aug
Jun
+20°
Jul
Mar
Aldebaran
PISCES
max elong west
2028 Aug 10
ORION
max elong east
2028 Mar 22
TAURUS
+10°
Feb
Betelgeuse
0°
Rigel
-10°

Tables

Planets compared

Here are the main orbital and physical characteristics of the eight major planets, also the Sun and the dwarf planet Pluto. The numbers are given to a fairly low level of exactness; some of them, such as orbital eccentricity, vary slightly over time. See the glossary for the meaning of terms such as "synodic." The columns are:

Mean distance from the Sun (semimajor axis of orbit), in astronomical units.

Orbital period, in years.

Synodic period, in days.

Eccentricity of orbit.

Inclination of orbit to the ecliptic, in degrees.

Equatorial diameter, in kilometers.

Mass, in Earth masses.

Flattening (excess of equatorial over polar diameter). When given as "0," it may exist but be smaller than 0.001.

Tilt of axis, in relation to the orbital plane (or, for the Sun, to the ecliptic plane), in degrees.

Sidereal rotation period, in days.

Known satellites.

	a au	Porb years	Psyn days	ecc	incl °	diameter km	mass Earths	flat	tilt °	rotat days	sat
Sun						1,391,400	333,000	0	7.3	25.05	
Mercury	0.39	0.24	116	.206	7.0	4,880	.055	0	2.0	59	
Venus	0.72	0.62	584	.007	3.4	12,104	.815	0	2.6	243	
Earth	1	1		.017		12,756	1	.003	23.4	1	1
Mars	1.52	1.88	780	.093	1.9	6,790	.107	.005	25.2	1.07	2
Jupiter	5.20	11.86	399	.049	1.3	142,980	317.8	.065	3.1	.41	67
Saturn	9.58	29.46	378	.057	2.5	120,540	95.2	.098	26 7	.44	82
Uranus	19.22	84.02	370	.046	.8	51,120	14.5	.023	97.8	.72	27
Neptune	30.11	165.8	367	.009	1.8	49,530	17.1	.017	28.3	.67	14
Pluto	39.48	247.9	367	.249	17.2	2,380	.187	.	122.5	6.39	5

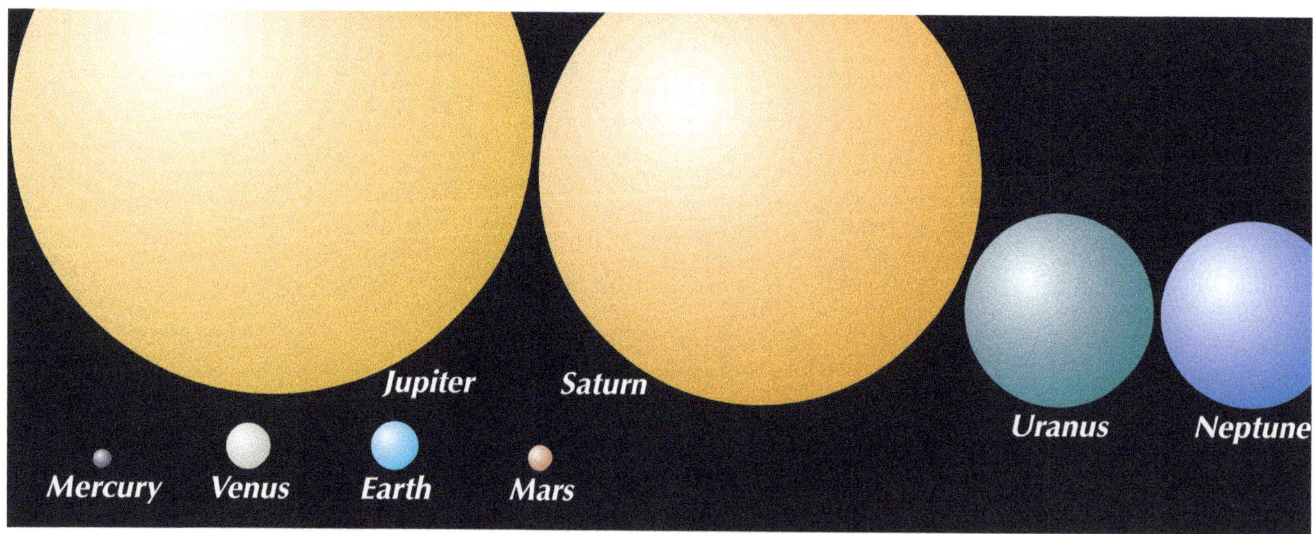

Venus's orbital elements

These numbers describe the orbit's size:
q, perihelion distance: 0.7184 astronomical units.
a, mean distance from the Sun, also called semimajor axis or radius vector: 0.7233 a.u.
A, aphelion distance: 0.7282 a.u.
P, orbital period: 0.6152 year (224.7 days).

The orbit's shape:
e, eccentricity, or difference from circular: 0.0068°.

The orbital plane's slope:
i, inclination to the ecliptic plane: 3.3946°.

The orbit's orientation:
Ω, longitude of the ascending node: angle from the vernal equinox to the ascending node: 76.69°.
ω, argument of perihelion: angle from the ascending node to the perihelion: 54.88°.

The timing between events as seen from Earth:
Synodic period, or average time between e.g. a superior conjunction to the next: 583.92 days (1.5987 years)

All of these numbers vary gradually over time, because of perturbations (the gravitational influence of the other moving bodies), so, unless given only roughly, they have to be given as for a certain epoch, such as (here) the beginning of the year 2000.

Five of these elements are needed for calculating an orbiting body's position at a date: e, i, Ω, ω, and either q or a; and a perihelion date (or a mean anomaly—angular distance from the vernal equinox—at a certain date). The three distance elements can be derived from each other by means of the eccentricity; and they are related to the period by Kepler's third law (period in years squared equals mean distance cubed). Though these elements describe the orbit and are used in calculations for comets and asteroids, in practice much longer formulae are used for the planets.

Venus's physical parameters

Radius: 6,051.8/6,378.1 km = 3,760.4 miles, diameter: 12,103.6 km = 7,520.8 miles. This is 0.95 of Earth's equatorial diameter.

Mass: 0.815 Earths.

Overall density: 5.24 grams per cubic centimeter. Cf. Earth's 5.51.

Tilt of rotational axis relative to orbital plane: 2.64°.

North pole points to right ascension 18h 11m (272.76°), declination 67.16°.

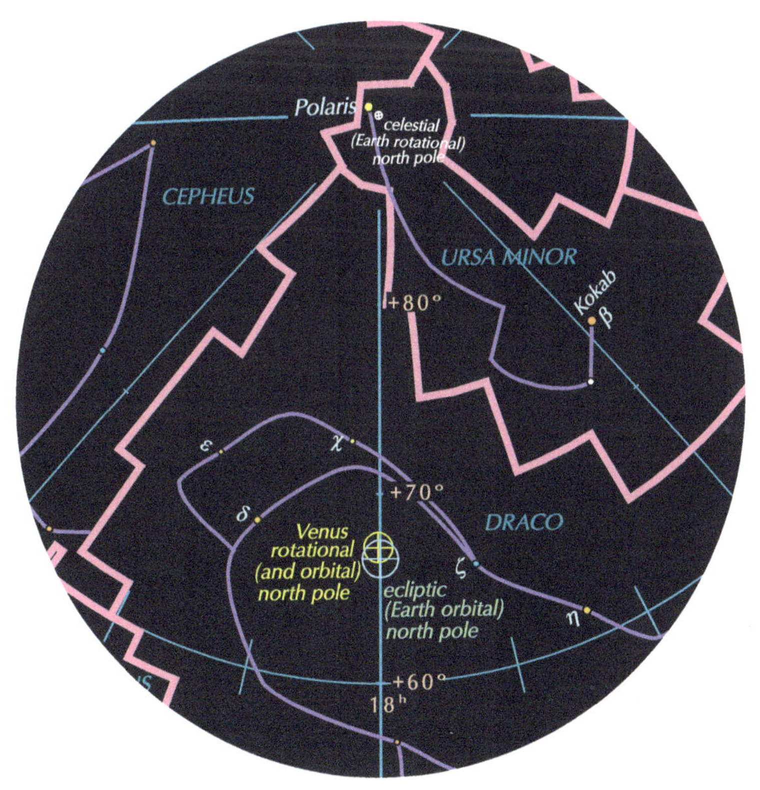

Venus's rotational north pole is in Draco, only 1.004° from the ecliptic north pole (the pole of Earth's orbital plane). This is because Venus's rotational pole is almost fixed, whereas Earth's describes a huge, age-long circle (radius 23.4°) around the ecliptic pole, because of precession. Venus's axial tilt to its orbital plane is only 2.6°, and (having no satellite like the Moon) it is scarcely affected by precession.

Glossary

Here are brief explanations of terms we have had to use. A fuller astronomical glossary, with pronunciations and etymologies, is our book *Albedo to Zodiac.*

albedo (Latin, "whiteness"): the fraction of incident light that a body reflects. Earth and Venus have albedos of 0.39 and 0.72, meaning that the visible cloud surface of Venus is much lighter.

aphelion: the outermost point in an orbit around the Sun. Cf. **periphelion**.

apoapsis: see **periapsis**.

apparent: as seen from Earth. The apparent diameter of a planet is the angle it subtends, as distinct from its physical width in kilometers or other linear measure.

apparition: a stretch of time during which a body is observable in the sky. E.g. for Venus, when it appears from behind the Sun, at or just after superior conjunction, to when it disappears in front of the Sun, at or before inferior conjunction (an evening apparition); or from inferior to superior conjunction (a morning apparition). For an outer planet or a star, the apparition begins in the dawn and ends in the evening sky.

appulse: instant when two bodies are nearest to each other, as viewed from Earth. Not quite the same as **conjunction**.

ascending node: the point where a planet's orbital plane crosses northward through the ecliptic.

astronomical unit (**a.u.** or **AU**): the mean distance between Sun and Earth.

axial tilt, or rotational inclination, of a planet is the angle its axis of rotation makes with the plane of its orbit.

azimuth: angular distance around or parallel to the horizon, usually measured from the north point eastward. So the azimuth of the east point, where the Sun rises at either equinox, is 90°. The horizon-based system of measurement, using azimuth and altitude, is called **altazimuth**.

celestial: of the sky. Celestial sphere: imaginary sphere around the Earth, on which celestial bodies and lines such as constellation boundaries can be drawn. Celestial equator: the rotational plane of the Earth; the series of points in the sky above Earth's equator.

conjunction of two bodies: instant when one is exactly north of the other, either in right ascension or in ecliptic longitude. Cf. **appulse**.

conjunction with the Sun: instant when a planet is in the same direction as the Sun, that is, has the same longitude, though it is usually north or south of the Sun. For Mercury and Venus, **inferior** conjunction is when they pass on the Sun's nearer side, and **superior** conjunction is on the farther side.

declination: angular distance north or south of the celestial equator, in the equatorial frame of reference; similar to latitude on the Earth.

descending node: the point where a planet's orbital plane crosses southward through the ecliptic.

dichotomy (from Greek *dikha*, "in two, asunder," and *tomia*, "a cutting"): the moment when a planet is seen to be bisected straight down the middle by the line between dark and light (the terminator).

direct motion: motion eastward against the starry background, the normal motion of planets in the solar system. Similarly, direct rotation is rotation such that the sky appears to move westward. Another word for the same thing is **prograde**; and the opposite kind of motion or rotation is **retrograde**.

eccentricity of an orbit: its departure from circularity. A circle has eccentricity 0; ellipses have eccentricity between 0 and 1; a parabola has eccentricity 1; a hyperbola has eccentricity more than 1.

ecliptic: the plane in which Earth revolves around the Sun; and the line on the sky marking this plane.

elongation: apparent angular distance of a body from the Sun.

epoch: the time for which something is intended to be true. For instance, the position (right ascension and declination) of a planet in 2020 may be given for the standard epoch of 2000, ignoring the change made by 20 years of precession.

equatorial system of mapping the sky and expressing the positions of objects: based on the celestial equator and the rotation of the Earth. It uses right ascension and declination, in contrast with the ecliptic system which uses longitude and latitude.

equinox: one of the two instants each year (about March 20 and September 23) when the Sun is on the celestial equator. Also, the two points on the map of the sky where this happens: the points where the ecliptic crosses the equator. The March or (northern) spring, or **vernal** equinox point, called for historical reasons the **First Point of Aries**, is the origin or zero-point for both the equatorial and the ecliptic systems of mapping the sky. Because of precession, all these points move back (westward) along the ecliptic. Cf. **solstice.**

evening sky: the half of the sky to the east of the Sun, that is, left of it as seen from the northern hemisphere.

First Point of Aries: see **equinox.**

flattening: the measure of how much a circle or sphere is compressed into an ellipse or ellipsoid. If the polar radius is 0.8 of the equatorial radius, the flattening is 0.2.

geocentric: from the viewpoint of the center of the Earth.

heliocentric: centered on the Sun.

inclination of an orbit: the angle at which it intersects the ecliptic.

inferior conjunction: passage of an inferior planet past the Sun, on the side nearer to Earth.

inferior planets: Mercury and Venus, which are nearer than us to the Sun (literally "lower," as the Latin word means, in the gravitational field of the Sun); which therefore have two kinds of conjunction (inferior and superior), never have oppositions, but instead are best seen near their greatest elongations.

invariant plane of the solar system: the general plane of its revolving matter, determined mainly by Jupiter. It differs by 2.65° from the ecliptic plane.

illuminated extent: the angular area of the visible sunlit part of the planet. The time of greatest illuminated extent may be of as much interest to observers with visual aid as the time of greatest elongation.

Kepler's laws: (1) A planet's orbit is an ellipse, with the Sun at one of the two foci. (2) A line from the planet to the Sun sweeps out an equal area in an equal time. (3) The orbital period (in years) squared equals the semi-major axis (average distance, in astronomical units) cubed.

latitude, ecliptic: angular distance north or south from the ecliptic.

limb: the visible edge of a spherical body, such as the Sun or Moon. Contrast **terminator.**

longitude: angular distance around, parallel to the ecliptic ; measured from the vernal equinox point.

magnitude: the astronomical system for expressing brightness; it started as a simple order of what were thought to be sizes. Very bright stars have magnitude 1 (a few, 0 or negative); stars just visible to the naked eye have magnitude around 6; Pluto is as dim as 14. The scale is actually logarithmic and each step of 1 represent a difference of about 2.5 in amount of light—to be precise, the factor is the 5th root of 100, or 2.5119 . . .

morning sky: the half of the sky to the west of the Sun.

node: see ascending and descending node.

periapsis and **apoapsis**: the nearest and farthest points in a body's orbit around a larger body. Cf. **perihelion** and **aphelion**, which are the cases where the larger body is the Sun. The apsis words are used to describe the orbits of spacecraft around other bodies, such as Venus.

perihelion: the innermost point in an orbit around the Sun. Cf. **aphelion.**

periodic comet: one that is in an orbit shorter than 200 years.

planet (Greek *planêtês*, "wanderer"): the 8 major planets are Mercury, Venus, Earth, Mars, Jupiter, Saturn, Uranus, Neptune. Dwarf planets, a category introduced in 2006, include Pluto and Ceres. Minor planets include asteroids and others.

precession: gradual change in the direction of a body's axis of rotation. For instance, Earth's north pole, over a period of 25,800 years, sweeps a circle in the sky with radius 23.4°, its axial tilt. Therefore the celestial equator slowly shifts. Since our sky maps are based on the equator and the vernal equinox point, all stars gradually change their map positions. Maps are now based on the standard epoch of 2000.

prograde motion or rotation: see **direct.**

radius vector of an orbit: the line from the Sun to the planet at the instant.

retrograde motion or rotation: backward, westward, the exception for the solar system. See **direct.**

revolve: to move in an orbit around another body. Contrast **rotate.**

right ascension (r.a.): angular distance parallel to the celestial equator, in the equatorial frame of reference; similar to longitude on the Earth. However, though it can be expressed in degrees, it is traditionally given in hours, minutes, and seconds. A point in the sky moves 15° westward in one hour of time.

rotate: to spin around an axis. Contrast **revolve.**

sidereal: of the stars (Latin *sidus*, "star," plural *sidera*) ; in the frame of reference outside the solar system.

sidereal period: the time taken by a planet to go once around the sky, as seen from the Sun, against the starry background.

solstice: one of the two instants each year when the Sun appears farthest north (about June 21) or south (about December 21) of the equator. Also, the two points on the map of the sky where this happens: where the ecliptic is at its maximum of about 23.4° from the celestial equator. Cf. **equinox.**

superior conjunction: passage of an inferior planet past the Sun, on the side farther from Earth.

synodic period: the time from an event of a planet as seen from Earth, such as its conjunction behind the Sun, to the next such event.

terminator: the line between the sunlit and dark sides of a planet. Contrast **limb**.

terrestrial: of the Earth (Latin *terra*). The terrestrial—Earth-like, or rocky—major planets are Mercury, Venus, Earth, and Mars.

vernal: of the spring season (Latin *ver*). See **equinox.**

zodiac: the band of constellations around the sky through which the Sun, Moon, and major planets pass. There are traditionally 12 zodiacal constellations (Pisces, Aries, Taurus, Gemini, Cancer, Leo, Virgo, Libra, Scorpius, Sagittarius, Capricornus, Aquarius), but the ecliptic also passes through Ophiuchus, and parts of some other constellations come near enough that the Moon and planets sometimes are in them.

It has been suggested that a book about Venus should include a glossary of erotic terms. No.

Shir ha-shirim asher li-Shlomo.
Yishshaqêni min-n'shiqoth pîhu, ki-tobim dodeka mi-yayin.
The song of songs, which is Solomon's.
Let him kiss me with the kisses of his mouth: for thy love is better than wine.

–Opening of the Song of Songs, also called the Song of Solomon, though its language shows that its anonymous author must have lived centuries later. (Li could mean "to" or "for" or "of.") It is the dialogue of a passionately and harmoniously loving couple, with a chorus of the daughters of Jerusalem. Though part of the Old Testament, it has no religious content, though some wish to think it allegorical.

If Venus provokes you to a mood for aphrodisiac reading, we might recommend the novels of Delver Maddingley (if they can still be found) and the cheerful fantasy entitled *Adventures of a Naked Girl*–all probably pseudonymous for good reason.

www.ingramcontent.com/pod-product-compliance
Lightning Source LLC
Chambersburg PA
CBHW041030050726
47599CB00018B/1918
9780934546812